MODERN METHODS IN ORCHID CONSERVATION:
THE ROLE OF PHYSIOLOGY, ECOLOGY AND MANAGEMENT

T0219715

MODERN METHODS IN ORCHID CONSERVATION: THE ROLE OF PHYSIOLOGY, ECOLOGY AND MANAGEMENT

H.W. PRITCHARD

Jodrell Laboratory
Royal Botanic Gardens, Kew
Wakehurst Place

The right of the
University of Cambridge
to print and sell
all manner of books
was granted by
Henry VIII in 1534.
The University has printed
and published continuously
since 1584.

CAMBRIDGE UNIVERSITY PRESS
Cambridge
New York   Port Chester
Melbourne   Sydney

Published by the Press Syndicate of the University of Cambridge
The Pitt Building, Trumpington Street, Cambridge CB2 1RP
40 West 20th Street, New York NY 10011, USA
10 Stamford Road, Oakleigh, Melbourne 3166, Australia

First published 1989

Britich Library cataloguing in publication data available

Library of Congress cataloguing in publication data available

ISBN 0 521 37294 1

Transferred to digital printing 2004

DQ

CONTENTS

PREFACE

This book is based on the proceedings of a national symposium on orchid conservation, which was held at the Royal Botanic Gardens, Kew, Richmond, Surrey, 12th & 13th November, 1986. It contains a series of articles on orchid conservation from three separate perspectives: in relation to physiology, ecology and management.

The intention of the symposium was to exchange viewpoints and to foster collaboration between scientists involved with experimental physiology and ecology, and members of the various national conservation organisations mainly concerned with management. The subject matter encompassed storage and germination of seeds and pollen, tissue culture, population biology, reserve and living collection management, and international trade regulations.

With this diversity of topics covered in this book it is hoped that it will be a useful starting point for those involved in all aspects of conservation, not just with orchids, providing an outline of the modern methods which are now available to the conservationist.

I would like to express my gratitude to all my colleagues at Wakehurst Place for their support and help in running the symposium. Thanks also to Mrs P. Bloomfield for secretarial services and Mrs J. Peschiera for help in preparing the artwork for the book. Finally, thanks to the contributors for their co-operation throughout, and to the staff of Cambridge University Press for their assistance in the production of this volume.

H.W. Pritchard

Jodrell Laboratory
Royal Botanic Gardens, Kew
Wakehurst Place

LIST OF CONTRIBUTORS

D. Butcher
Micropropagation Unit, Royal Botanic Gardens, Kew, Richmond, Surrey, TW9 3AB.

R. Cox
The Institute of Terrestrial Ecology, Monks Wood Experimental Station, Abbots Ripton, Huntingdon, PE17 2LS.

L. Farrell
Nature Conservancy Council, Northminster House, Peterborough, PE1 1UA.

R. Fitzgerald
Nature Conservancy Council, Church Street, Wye, Ashford, Kent, TN25 5BW.

G. Hadley
Department of Plant Science, University of Aberdeen, Aberdeen, AB9 2UD.

N.S.J. Hailes
School of Applied Sciences, The Polytechnic, Wolverhampton, Wulfruna Street, Wolverhampton, WV1 1SB.

M.J. Hutchings
School of Biological Sciences, University of Sussex, Falmer, Brighton, BN1 9QG.

S.G. Knees
Economic and Conservation Section, Royal Botanic Gardens, Kew, Richmond, Surrey, TW9 3AB.

S.A. Marlow
Micropropagation Unit, Royal Botanic Gardens, Kew, Richmond, Surrey, TW9 3AB.

H.J. Muir
Sainsbury Orchid Project, Royal Botanic Gardens, Kew, Richmond, Surrey, TW9 3AB.

G.F. Pegg
Department of Horticulture, University of Reading, Earley Gate, Reading, RG6 2AT.

F.G. Prendergast
Jodrell Laboratory, Royal Botanic Gardens, Kew, Wakehurst Place, Ardingly, West Sussex, RH17 6TN.

H.W. Pritchard
Jodrell Laboratory, Royal Botanic Gardens, Kew, Wakehurst Place, Ardingly, West Sussex, RH17 6TN.

P.T. Seaton
Kidderminster College of Further Education, Hoo Road, Kidderminster, Worcestershire, DY10 1LX.

J. Stewart
Sainsbury Orchid Project, Royal Botanic Gardens, Kew, Richmond, Surrey, TW9 3AB.

S. Tasker
Orchid Unit, Royal Botanic Gardens, Kew, Richmond, Surrey, TW9 3AB.

S. Waite
School of Biotechnology, The Polytechnic of Central London, 115 New Cavendish Street, London W1M 8JS.

R.C. Warren
Equatorial Plant Company, 73 Dundas Street, Edinburgh, EH3 6RS.

T.C.E. Wells
The Institute of Terrestrial Ecology, Monks Wood Experimental Station, Abbots Ripton, Huntingdon, PE17 2LS.

J.J. Wood
The Herbarium, Royal Botanic Gardens, Kew, Richmond, Surrey, TW9 3AB.

H.W. PRITCHARD AND F.G. PRENDERGAST

Factors influencing the germination and storage characteristics of orchid pollen

Introduction

As an adjunct to seed storage for genome preservation, orchid pollen storage has much to offer: to the hybridist wishing to overcome flowering asynchrony in species and/or to introduce wild genome contributions into cultivated taxa; to the gene bank manager seeking to preserve a large quantity of genetic material in a small facility; and to the conservationist anxious to preserve species, even if this can only be achieved through the pollination of plants held *ex situ* in orchid collections with stored pollen. Pollen storage may also be of considerable future importance to the biotechnologists, if the process of haploid plant production via embryo development from pollen grains (i.e. androgenesis) can be extended to orchids.

The literature on orchid pollen germination is surprisingly limited with previous studies concentrating on optimising the composition of the germination medium, particularly the sugar level. Pfundt (1910), working with two European species and a range of sugar levels from 5-20%, observed a preference for 5-10% sugar. Miwa (1937) similarly found orchid pollen generally germinates best when using cane sugar at around 3-10%. Although more exacting studies on the optimal chemical composition of the germination medium have been performed, particularly with reference to plant hormones (Curtis & Duncan 1947; Rao & Chin 1972), sugar level remains the most important single chemical factor in stimulating orchid pollen germination on artificial medium; its action being osmotic, preventing grain bursting whilst avoiding plasmolysis, rather than as a heterotrophic source of carbon (see Stanley & Linskens 1974). Stigmatic extract or fluid does not consistently stimulate germination above the level obtained with an optimal sucrose medium (Curtis & Duncan 1947; Rao & Chin 1972).

Like germination, investigations into the storage of orchid pollen are few. Air-dry storage at 45 °F (7 °C) has proved successful for up to 12 months with two species and one hybrid *Dendrobium* (Meeyot & Kamemoto 1969). Similar success with other species of *Dendrobium*, *Vanda*, *Cymbidium* and *Arachnis* has been achieved using air-dry storage at 4-6 °C for a maximum of 280 days (Shijun 1984). Pfundt (1910) showed that the longevity of *Listera ovata* pollen can be improved using very low air relative humidity, achieved by storage over sulphuric acid (≡ 0.005% r.h.). However,

Table 1. *Species used throughout the study and medium sucrose level used in experiments other than those presented in Figure 1.*

Species	Sucrose level %
Anacamptis pyramidalis (L.) L.C.M. Richard	1
Cymbidium elegans Lindl.	3
C. tracyanum L. Castle	1
Dactylorhiza fuchsii (Druce) Soó	1
D. maculata (L.) Soó	1
Epipactis purpurata Sm.	1
Gymnadenia conopsea (L.).R. Br.	1
Listera ovata (L.) R. Br.	1
Orchis mascula (L.) L.	1
O. morio L.	1
Spiranthes spiralis (L.) Chevall.	3

the use of very dry storage is not advisable for all pollen. For example, Gramineae pollen is generally intolerant of desiccation (see References in Barnabas 1985). Moreover, reports on the effect of short-term drying and long-term storage over desiccants in orchid pollen are conflicting. Whereas *Listera ovata* (Pfundt 1910), *Cattleya mossiae* (Curtis & Duncan 1947) and *Dendrobium* 'Lady Hamilton' (Ito 1965) appear relatively tolerant of drying, such conditions are harmful in other *Dendrobiums* and *Oncidium stipitatum* (Meeyot & Kamemoto 1969). Thus, no firm conclusions can be drawn concerning the relationship between longevity and the relative humidity of the storage environment, particularly as the actual pollen moisture content was undetermined in these studies.

Ito (1965) pioneered the sub-zero storage of orchid pollen, observing that *Dendrobium nobile*, *Dendrobium* 'Lady Hamilton' and *Calanthe furcata* pollen germinated 'well' after 957 days at -79 °C. Moreover, *Dendrobium nobile* pollen survived 93 days at -79 °C in the presence of a chemical cryoprotectant (glycerol/ethyl alcohol mixture). Such chemicals are commonly used in the routine storage of hydrated plant tissues at sub-zero temperatures (Withers 1980).

Previous investigations have been mainly concerned with commercially important orchid species, mostly of tropical origin. Only Molisch (1893) and Pfundt (1910) have, briefly, considered European species. Our studies have dealt mostly with European species of British provenance (Table 1) and report not only the factors affecting *in vitro* germination but also storage effects, including cryopreservation and pattern of viability loss during storage. Such information should be useful in the future development of a method for the conservation of orchid species through pollen gene banks, whether on a small, individual level or on a large, national or international scale.

Materials and methods

Source of material

Fresh pollen was collected under licence from sites in Sussex, or provided from plants in the Living Orchid Collection, Kew. Whole pollinaria (i.e. viscidium, caudicle and pollinium combined) were removed for experimental use. However, due to the limited availability of pollen from some species, application of all treatments to the full range of species listed in Table 1 was not possible.

In vitro *germination*

Freshly collected and stored pollinaria were sown on a sterile basal medium of 100 mg dm^{-3} boric acid, 1% agar, pH 5.6 and sucrose levels between 1-10% (w/v). Three to five pollinaria per treatment were incubated under 'warm white' fluorescent light (14 μmol m^{-2} s^{-1}, 12 h photoperiod) at 26 °C for 24 h or 48 h. For scoring, pollinaria were placed in a drop of distilled water and the pollen teased out. Fifty tetrads per pollinarium (i.e. 200 grains) were scored under the microscope: only grains with pollen tubes longer than the grain diameter were counted as germinated.

Desiccation and 'conditioning'

For desiccation, pollinaria, in open Petri dishes, were placed above silica gel in a sealed container for up to three days at 21 °C. 'Conditioning' was performed by placing the pollinaria in a saturated atmosphere above water in a sealed container for one day at 21 °C.

Storage

Whole pollinaria were stored for up to one year under the following conditions:

1. At 2 °C, in unsealed Petri dishes.
2. At 2 °C, on 1% agar in unsealed Petri dishes.
3. At -20 °C, in 2 cm^3 capacity screw-cap polypropylene ampoules.
4. At -196 °C, as detailed in 3 above.
5. At -20 °C in 1 cm^3 of cryoprotectant (either 0.5 M glycerol or dimethyl sulfoxide (DMSO)) inside polypropylene ampoules.
6. At -196°C, as detailed in 5 above.

Pollinaria were cryoprotected by incubation in the solution for 45 minutes at 2 °C prior to storage. Uncryoprotected pollinaria were cooled to -20 °C and -196 °C by direct transfer of the ampoules into a chest freezer or liquid nitrogen, respectively. This also applied to cryoprotected pollinaria stored at -20 °C. For cooling cryoprotected pollinaria to -196 °C, ampoules were initially suspended 16 cm above the surface of liquid nitrogen in an open-neck Dewar flask. By altering the height of the ampoule holder, controlled rates of cooling to -35 °C could be achieved, specific details of which are given with each experiment reported. During cooling, the ampoule temperature was monitored using a Ni/Cr thermocouple and electronic thermometer. Ampoules stored at -20 °C were rewarmed at room temperature; those stored at -196 °C, by immersion in a 40 °C water bath.

At least three pollinaria were withdrawn for germination testing at specified intervals during the storage period, and the container returned to the appropriate temperature following sampling. Following storage, cryoprotected pollinaria were plated directly on to germination medium without washing out of the cryoprotectant.

Relative humidity and moisture content determination

The relative humidity in the 2 °C storage incubator was monitored by humidity sensors (Humisensor HTPC 111M, Ancom Ltd., Cheltenham, U.K.) at various times during the storage experiments.

For gravimetric moisture content determination, at least four pollinia (i.e. pollinaria with caudicle and viscidium removed) were dried for 17 h in a fan-driven oven at 103 ± 2 °C (International Seed Testing Association 1985). All weighings were performed using a Sartorius ultramicro balance (seven decimal places) and water contents were calculated on a fresh weight basis.

Results and discussion

Factors influencing the level of germination

Length of germination test. In the species studied here pollen germination is protracted to the extent that a germination test time of 24 h to 48 h is necessary. This concurs with previous studies (Miwa 1937; Curtis & Duncan 1947; Rao & Chin 1972). Although a 72 h testing period has been used with other orchid species (Miwa 1937; Meeyot & Kamemoto 1969) we found this led to fungal contamination of the medium. Orchid grains are bicellular i.e. the generative cell has not divided (see Wirth & Withner 1959) and such grains are known to germinate slowly, generally over a period of hours (Cerceau-Larrival & Challe 1986). In contrast, germination is generally initiated within a few minutes in tricellular grains i.e. the generative cell has already divided (Hoekstra & Bruinsma 1978).

Medium sucrose level. The relationship between *in vitro* germination and sucrose level for pollen of eight species is presented in Figure 1. Two species, *Dactylorhiza maculata* and *Listera ovata*, exhibit little preference for sucrose between 1-10%

(Figure 1A). However, the other six species show higher germination on either 1 or 3% sucrose (Figure 1B, C & D); although in *Cymbidium elegans* germination is not significantly higher at 3% than at 5% sucrose (Figure 1C). In most of these species plasmolysis of grains frequently occurs at the 10% sucrose level. Where germination is equally high at two or more sucrose levels the lower was used in all subsequent experiments.

The preference for low sucrose levels in the germination medium (see also Table 1) is in general agreement with previous work on orchid pollen (Molisch 1893; Pfundt 1910; Miwa 1937; Curtis & Duncan 1947; Meeyot & Kamemoto 1969; Rao & Chin 1972), indicating low internal grain osmotic potentials. These appear much lower than for a majority of angiosperm pollen which, both as compound grains and as monads,

Figure 1. Effects of medium sucrose level on orchid pollen germination. A, *D. maculata* (●), *L. ovata* (O). B, *D. fuchsii* (●), *G. conopsea* (O). C, *S. spiralis* (●), *C. elegans* (O). D, *A. pyramidalis* (●), *C. tracyanum* (O). Data represents the mean ± S.D. of at least three pollinaria, with 50 tetrads scored from each after incubation on medium at 26 °C for 24 h.

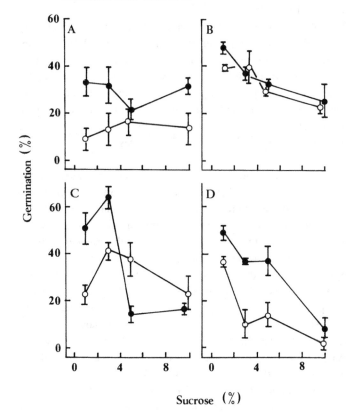

Table 2. *Equilibrium moisture contents (e.m.c.) for orchid pollen after 24 hours over silica gel desiccant at 21 °C, and during longer-term storage at 85% relative humidity and 2 °C.*

Species	Silica gel, 21 °C	85% r.h., 2 °C
Spiranthes spiralis	6.3 ± 0.5	14.9 ± 2.3
Listera ovata	7.5 ± 1.6	13.5 ± 1.3
Dactylorhiza maculata	6.5 ± 0.6	17.6 ± 1.0
Orchis mascula	6.6 ± 0.5	18.3 ± 3.9
Cymbidium elegans	6.4 ± 0.2	18.4 ± 0.2
Orchis morio	5.8 ± 0.2	19.1 ± 2.1
Dactylorhiza fuchsii	6.0 ± 1.7	20.3 ± 2.9

Data represents the mean ± S.D. for at least three pollinia

generally germinate best at sucrose levels between 10–30% (Rao & Ong 1972; Cerceau-Larrival & Challe 1986).

The level of orchid pollen germination obtained on artificial medium is considerably higher than that observed in other compound grains (Rao & Ong 1972). Medium composition, particularly sucrose level, influences the germination response. However, in some species germination is highly variable even at one sucrose level (as shown by the standard deviation bars in Figure 1), suggesting factors other than medium composition may also be important. One possible contributor to this variation in germination response currently being investigated is the occurrence of heterogeneous populations of pollen on individual inflorescences.

Influence of desiccation. The beneficial effects of reduced moisture for the maintenance of pollen viability during storage has been reported for numerous angiosperm species (for review see Stanley & Linskens 1974). Thus as a prelude to storage, the effects of drying over silica gel desiccant for up to three days on germination was investigated. Drying over silica gel for 24 h reduces pollen moisture content to less than 8% (Table 2) and in seven of the eight species studied, drastically reduces germinability to around 10% or lower (Figure 2). Only *Orchis mascula* exhibits some desiccation tolerance: 40% viable after one day's drying (Figure 2B).

One method of improving the germination response in dehydrated grains is by limiting the initial water uptake to the vapour phase. This ensures slow rehydration and successful reconstitution of the phospholipids in the dry plasma membrane into a lamellar pattern thus reducing the efflux of cytoplasmic solutes when the pollen is placed on the germination medium (Shivanna & Heslop-Harrison 1981). This technique, known as 'conditioning', is partially effective in only two of the six orchid species studied, and even here the treatment did not completely restore the germinability to its initial level (Figure 2). Perhaps, as with maize (Barnabas 1985) and rye (Shivanna & Heslop-Harrison 1981), this lack of response to 'conditioning' is due to irreversible membrane changes resulting from excessive dehydration.

The results presented here for eight orchid species support the observations of long-term desiccation sensitivity in two species and one hybrid of *Dendrobium* (Meeyot & Kamemoto 1969), but conflicts with reports of relative desiccation tolerance in three

Figure 2. Influence of desiccation on orchid pollen germination. A, *S. spiralis.* B, *O. mascula.* C, *O. morio* D, *A. pyramidalis.* E, *G. conopsea.* F, *D. fuchsii.* G, *D. maculata.* H, *L. ovata.* Pollinaria were dried over silica gel for up to three days (▨; d1, d2, d3), and subsequently 'conditioned' for one day (▨; +c) at 21 °C. Data as Figure 1, except 48 h incubation in germination test.

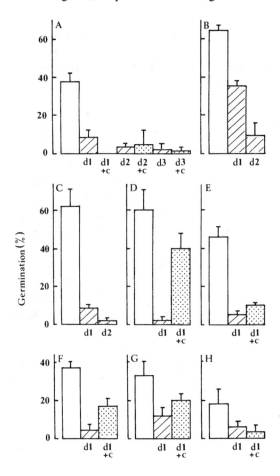

Drying/conditioning treatments

other orchid species (Pfundt 1910; Curtis & Duncan 1947; Ito 1965). Clearly, if drying is to become a routine part of orchid pollen gene banking, much shorter drying periods may have to be considered for some species. The equilibration of orchid pollen with pre-determined optimal relative humidities prior to storage so avoiding excessive dehydration appears more practical.

The variation in response between species to the factors investigated which influence the *in vitro* germination of orchid pollen suggest that before a pollen gene bank can operate using a limited number of artificial germination media and a unified pollen drying strategy, fundamental studies are required on a much broader range of species than hitherto considered.

Storage characteristics

Refrigerated storage at high relative humidity and in an hydrated state. Fully hydrated storage on 1% agar at 2 °C results in a rapid loss in viability with little viability remaining after six weeks' storage (Figure 3A). The shortest lived pollen of the five species studied is *Spiranthes spiralis*.

An improvement in longevity is observed when storing pollen of eight species at 2 °C in equilibration with the prevailing atmospheric moisture: 85 ± 6% relative humidity. *Dactylorhiza fuchsii* (Figure 3B), *Orchis morio*, *O. mascula* and *Anacamptis pyramidalis* (Figure 3C) retain at least 50% of their initial viability after 60 days storage. In contrast, the other four species studied lose at least 50% of their viability within 40 days of storage: *Spiranthes spiralis* again showing the poorest longevity (Figure 3C). This difference between species longevity could be due to differential pollen moisture contents during storage. As Table 2 shows, the equilibration moisture contents (e.m.c.) for pollen in 85% relative humidity do indeed vary considerably between species, from 15% to 20%. As with angiosperm seeds this is likely to be a consequence of the chemical composition of the pollen, particularly in relation to the oil or fat content, which is known to vary between species (see Stanley & Linskens 1974). However, the difference between the viability loss curves of separate species which do not differ significantly in moisture content (e.g. *Cymbidium elegans* and *Orchis mascula*, Figure 3C, Table 2), refutes a correlation between longevity and moisture content alone: longevity is also intimately related to species.

Another feature in six of the eight species studied is that the viability loss lines are convex when plotted on a probability scale: there is a pronounced 'shoulder' followed by a rapid loss in viability (Figure 3B & C). This contrasts with the straight lines which are produced when transforming the data available for pollen of seven other angiosperm species: *Elaeis guineensis* (Ekaratne & Senathirajah 1983), *Moringa oleifera* (Singh *et al.* 1983), *Gallardia arisata* (Panchaksharappa & Tirlapur 1984), blackberry (Perry & Moore 1985), pecan (Wetzstein & Sparks 1985), 'Ginyose' chestnut (Akihama & Omura 1986) and *Sparmannia africana* (Cerceau-Larrival & Challe 1986) (data not presented). Therefore, orchid pollen survival curves cannot be described as conforming to a negative cumulative normal distribution: very many more grains in the population only survive for short periods and many fewer survive for long periods than expected. Interestingly, transformation of Meeyot & Kamemoto's (1969) data reveals that undried *Dendrobium* pollen also exhibits convex survival curves. We believe the convex shape of the survival curves observed in this study could relate to four different, but possibly interacting, processes.

Figure 3. Survival curves for orchid pollen stored at 2 °C. A, Pollen stored on 1% agar. *O. mascula* (●), *D. fuchsii* (○), *O. morio* (■), *C. elegans* (□), *S. spiralis* (▲), B & C, Pollen stored in equilibration with 85% r.h. B, *D. fuchsii* (●), *L. ovata* (○), *E. purpurata* (■). C, *A. pyramidalis* (●), *O. morio* (○), *O. mascula* (■), *C. elegans* (□), *S. spiralis* (▲). Three pollinaria withdrawn per sample point and 50 tetrads scored from each. Germination conditions as Figure 2.

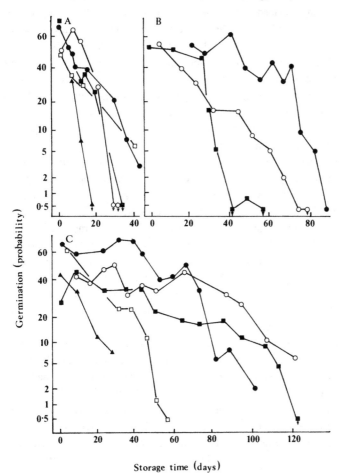

Firstly, there is a negative interaction between live and dead grains within an individual tetrad. Perhaps this occurs through the accumulation of secondary metabolic products, such as organic acids, in dying/dead grains, and their subsequent permeation to other grains in the tetrad. Given that the rapid phase of viability loss occurs at around 40% viability (Figure 3B & C), this effect presumably develops when only one or two grains within a tetrad remain viable.

Secondly, and in the same context as above, an accumulation of metabolic products during storage may alter the internal grain water potential sufficiently that higher levels of sugar are required in the artificial germination medium (Vasil 1962). However, if this is a common feature of pollen viability loss it is surprising that curved viability loss lines are not a frequent feature in non-orchidaceous species as well.

Thirdly, viability loss during early storage is imperceptible in a number of species because the loss is being replaced by the concurrent maturation of other tetrads within the population. Implicit in this is that the relatively low pollen viabilities generally recorded in this study can be attributed to the initial physiological incompetence for germination of part of the population. This phenomenon has been observed in *Zea mays* pollen, (Pfahler & Linskens 1973) and may reflect a requirement for moderate drying to achieve grain maturation, as seen with lily pollen (Lin & Dickinson 1984). In this regard, the use of a cool storage temperature and high relative humidity in the present study may account for the apparently protracted maturation of orchid pollen. Studies on orchid pollen development and maturation are in progress.

Fourth and finally, the occurrence of convex viability loss lines may directly relate to the pollen moisture content during storage. Note that the two species which appear to have a linear relationship between germination and storage time, *L. ovata* and *S. spiralis* (Figure 3B & C), have the lowest storage moisture contents: in both cases below 15% (Table 2). Like lettuce seeds when stored above 15% moisture content (Ibrahim & Roberts 1983), orchid pollen too exhibits a skewed normal distribution for the frequency of grain death within the population. In the case of lettuce seeds, this shape is attributed to the activation of biochemical repair mechanisms at moderate moisture levels. Such processes which benefit from the presence of oxygen (Ibrahim *et al*. 1983) would not be hindered in our experiments as the pollen was stored in high relative humidity in unsealed Petri dishes at 2 °C.

Our data, together with the examples cited in the text, do generally support an explanation for convex survival curves based on pollen moisture content. Pollen of the six angiosperm species referred to earlier as having linear survival curves had been dried prior to or during storage; although such a line is observed in one treatment of undried oil palm pollen (Ekaratne & Senathirajah 1983). On the other hand, convex pollen survival curves generally occur with undried, or at least relatively moist, orchid (Figure 3) and jojoba pollen (transformed data of Lee *et al.* 1985, data not presented). However, transformation of Rushton's (1977) data for *Quercus* pollen stored above silica gel also results in such a curve. Future resolution of this concept will depend on experiments in which pollen is stored at a range of pre-determined moisture contents.

Although we believe the fourth explanation is of greatest importance, we cannot at this time exclude the possibility of a confounding interaction with any one of the other three proposals to explain the occurrence of convex survival curves in orchid pollen. To remedy this, further experiments should investigate: individual grain deaths within a tetrad at the biochemical and ultrastructural level; medium composition adjustment for use with aged grains; influence of desiccation on grain maturation; and the shape of pollen survival curves in relation to a range of moisture contents.

Cryopreservation at −196 °C. Cryopreservation of pollen has been successfully applied to a range of angiosperm species, increasing the longevity of pollen compared to storage at refrigerated temperatures (Sedgley 1981; Haunold & Stanwood 1985; Lee *et al.* 1985). In general, pollen is stored in an air-dry state, or after a period of drying, either over desiccant or after freeze-drying (see Akihama & Omura 1986). Unusually, Ito (1965), using *Dendrobium* and *Calanthe* pollen, successfully used a cryoprotectant (a mixture of glycerol and ethylalcohol) with storage at −79 °C. We further investigated the effects of two commonly used cryoprotectants, glycerol and DMSO, on the viability of orchid pollen in association with cooling to liquid nitrogen temperatures (−196 °C).

Incubation of pollen of *C. elegans* (Table 3A) and *O. mascula* (Table 3B) in cryoprotectants does not significantly reduce pollen viability, except in *O. mascula* with 1M glycerol. Thus, cytotoxicity of the cryoprotectants is generally not apparent. Interestingly, however, treatment with both 0.5 M and 1 M glycerol increases tube lengths during subsequent *in vitro* germination in *C. elegans* (Table 3A). It remains unclear whether this stimulation in tube elongation results from glycerol acting as a source of carbohydrate for heterotrophic pollen tube growth or wall formation.

The effects of subsequent exposure of cryoprotected *C. elegans* and *O. mascula* pollen to slow cooling to −35 °C followed by direct transfer to liquid nitrogen, storage for 1 h and rapid rewarming are also presented in Table 3. Apart from *C. elegans* pollen in DMSO, cryopreservation treatments result in a significant decrease in pollen viability compared with the controls, although the promotory effects of glycerol on tube elongation remain (Table 3A). Three of five more species studied show no adverse effects of cryoprotection in 0.5 M DMSO, but do exhibit reduced viability after subsequent cryopreservation (Table 4). Only in *D. maculata* and *L. ovata* is germination not significantly reduced (P > 0.05) following cryopreservation in the presence of cryoprotectant (Table 4). Although cryoprotection is a prerequisite to successful cryopreservation of hydrated plant tissues (Withers 1980), obviating the need for a pre-drying phase, the low level of success reported here – in stark contrast to Ito's (1965) findings – prompted alternative methods of cryopreservation to be

Table 3. *Effects of cryoprotection and cryopreservation on orchid pollen germination and tube length. A, C. elegans. B, O. mascula. Cryopreservation: ampoules cooled at either 3.7 °C min^{-1} (A) or 2.1°C min^{-1} (B) to –35 °C, direct transfer to liquid nitrogen, storage for 1 h and rapid warming in a water bath at 40 °C. Data is mean ± S.D. for at least three pollinaria germinated at 26 °C for 48 h. Significant differences (P < 0.05) between treatments denoted: a, cryoprotected versus control; b, cryopreserved versus cryoprotected; c, cryopreserved versus control.*

Cryoprotectant	Cryoprotected	Cryopreserved
	Germination (%)	
A.		
Control	51.3 ± 6.7	
0.5M DMSO	50.2 ± 2.6	40.0 ± 1.1[b]
1.0M DMSO	44.2 ± 5.5	37.6 ± 6.4
0.5M Glycerol	53.8 ± 4.9	30.0 ± 5.8[bc]
1.0M Glycerol	36.2 ± 5.0	18.7 ± 2.2[bc]
	Tube length (μm)	
Control	91.3 ± 5.3	
0.5M DMSO	99.7 ± 8.0	72.3 ± 1.1[bc]
1.0M DMSO	123.3 ± 16.2	118.8 ± 20.6
0.5M Glycerol	145.3 ± 19.8[a]	150.0 ± 23.3[c]
1.0M Glycerol	169.2 ± 10.0[a]	159.0 ± 23.2[c]
	Germination (%)	
B.		
Control	32.8 ± 5.4	
0.5M DMSO	21.7 ± 5.6	12.3 ± 1.1[c]
1.0M DMSO	27.5 ± 8.1	11.8 ± 3.2[c]
0.5M Glycerol	27.8 ± 3.4	8.2 ± 2.6[bc]
1.0M Glycerol	20.6 ± 2.6[a]	3.0 ± 1.7[bc]

considered. Thus in addition, storage experiments with air-dry pollen (about 15% moisture content) were performed at both –20 °C and –196 °C. These results are presented in Figure 4.

In the two species studied, *Dactylorhiza fuchsii* and *Anacamptis pyramidalis*, there is little loss in viability during 12 months air-dry storage at –20 °C or –196 °C (Figure

4). Thus, unlike in avocado pollen (Sedgley 1981), conventional subzero storage considerably improves orchid pollen longevity when compared with storage at refrigerated temperatures (cf. Figure 3 & 4). Moreover, there is a stepwise reduction in viability in the cryoprotected pollen stored at both $-20\,°C$ and $-196\,°C$ on successive withdrawal of the same ampoule for sub-lot viability testing. This emphasises the damaging effects of cryopreservation in the presence of cryoprotectant

Table 4. *Effect of cryoprotection with 0.5M DMSO and cryopreservation on orchid pollen germination. Cryopreservation: initial cooling rate 2.1 °C min⁻¹. All other details as Table 3.*

Species	Germination (%)		
	Control	Cryoprotected	Cryopreserved
Dactylorhiza maculata	31.1 ± 4.5	34.7 ± 8.1	40.5 ± 6.1
Listera ovata	25.2 ± 4.5	29.6 ± 11.9	23.4 ± 5.0
Dactylorhiza fuchsii	58.4 ± 3.9	58.7 ± 1.7	22.9 ± 6.5[bc]
Anacamptis pyramidalis	41.9 ± 10.9	26.6 ± 10.2	0.9 ± 3.0[bc]
Gymnadenia conopsea	25.9 ± 4.3	3.7 ± 4.8[a]	0.9 ± 1.4[bc]

Figure 4. Effects of cryopreservation and storage period on *D. fuchsii* (A) and *A. pyramidalis* (B) pollen viability. Storage at $-20\,°C$ (●,○) or $-196\,°C$ (■, □). Storage ampoules contain air (closed symbols) or 0.5M DMSO (open symbols). Cooling rate for $-20\,°C$ was 2.5 °C min⁻¹ and for $-196\,°C$ storage with cryoprotectant was as Table 4. Data is mean ± S.D; germination details as Figure 3.

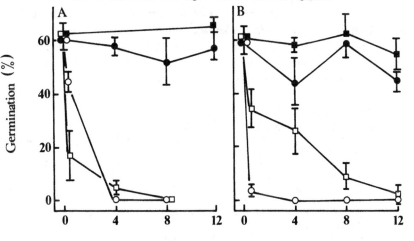

Germination (%)

Storage time (months)

previously shown in Tables 3 & 4. In contrast, in our experience *D. fuchsii* pollen in an air-dry state survives 12 cooling/rewarming cycles with –20 °C storage (cf. 41% control germination with 51% treatment germination; n = 3 × 50 tetrads), and *O. mascula* pollen tolerates 10 temperature cycles to –196 °C (cf. 51% control with 46% treated; n = 3 × 50 tetrads).

Conclusions

Germination of orchid pollen, like that of other bicellular pollen, is slow. The level of germination obtained *in vitro* depends partly on the medium sucrose content: low levels are preferred. Desiccation over silica gel reduces pollen viability, probably through excessive dehydration. Thus, equilibration of pollen prior to storage with constant relative humidity solutions to give pre-determined pollen moisture contents, should be considered in future.

The storage characteristics of orchid pollen indicate that at refrigerated temperatures and in equilibrium with high relative humidity the pollen is quite long lived (i.e. up to three months), but generally has unusual, convex survival curves. Four possible contributory factors to such curves have been suggested: interaction between live and dead grains in a tetrad; altered medium requirement of aged grains; physiological immaturity of grains; and relatively high grain moisture contents. Future research should be directed towards evaluating the role of each factor.

Long-term storage of pollen in air-filled ampoules at sub-zero (°C) temperatures is better than storage in the presence of cryoprotectants. Such storage conditions offer the potential to extend the longevity of orchid pollen to at least one year.

Acknowledgements

The assistance of the following people/organisations in securing pollen supplies for the project is gratefully acknowledged: C. Bailes, G. Barter, J. Harmer, Nature Conservancy Council, T. Schilling, and S. Tasker. Thanks to Mr K.R. Manger for technical assistance, Mr R.D. Smith for commenting on the manuscript and to Mrs P. Bloomfield for providing secretarial services.

References

Akihama, T. & Omura, M. (1986). Preservation of fruit tree pollen. In *Biotechnology in Agriculture and Forestry*, Vol. 1, Trees 1, ed. Y.P.S. Bajaj, pp. 101–12. Heidelburg: Springer–Verlag.

Barnabas, B. (1985). Effect of water loss on germination ability of maize (*Zea mays* L.) pollen. *Ann. Bot.*, **55**, 201–4.

Cerceau-Larrival, M-T. & Challe, J. (1986). Biopalynology and maintenance of germination capacity of stored pollen in some angiosperm families. In *Pollen and Spores: Form and Function*, ed. S. Blackmore & I.K. Ferguson, pp. 152–64. London: Academic Press.

Curtis, J.T. & Duncan, R.E. (1947). Studies on the germination of orchid pollen. *Am. Orchid Soc. Bull.*, **16**, 594–7, 616–9.

Ekaratne, S.N.R. & Senathirajah, S. (1983). Viability and storage of pollen of the oil palm, *Elaeis guineensis*. Jacq. *Ann. Bot.*, **51**, 661–8.

Haunold, A. & Stanwood, P.C. (1985). Long–term preservation of hop pollen in liquid nitrogen. *Crop Sci.*, **25**, 194–6.

Hoekstra, F.A. & Bruinsma, J. (1978). Reduced independence of the male gametophyte in angiosperm evolution. *Ann. Bot.*, **42**, 759–62.

Ibrahim, A.E. & Roberts, E.H. (1983). Viability of lettuce seeds. I. Survival in hermetic storage. *J. Exp. Bot.*, **34**, 620–30.

Ibrahim, A.E., Roberts, E.H. & Murdoch, A.J. (1983). Viability of lettuce seeds. II. Survival and oxygen uptake in osmotically controlled storage. *J . Exp. Bot.*, **34**, 631–40.

International Seed Testing Association (1985). International rules for seed testing 1985. Determination of moisture content. *Seed Sci. Technol.*, **13**, 338–41.

Ito, I. (1965). Ultra-low temperature storage of pollinia and seeds of orchids. *Japan Orchid Soc. Bull.*, **11**, 4–15.

Lee, C.W., Thomas, J.C. & Buchmann, S.L. (1985). Factors affecting *in vitro* germination and storage of jojoba pollen. *J. Am. Soc. Hortic. Sci.*, **110**, 671–6.

Lin, J-J. & Dickinson, D.B. (1984). Ability of pollen to germinate prior to anthesis and effect of desiccation on germination. *Plant Physiol.*, **74**, 746–8.

Meeyot, W. & Kamemoto, H. (1969). Studies on storage of orchid pollen. *Am. Orchid Soc. Bull.*, **38**, 388–93.

Miwa, A. (1937). Test of the germinating power of orchid pollen. *Orchid Rev.*, **45**, 345–9.

Molische, H. (1893). Zur Physiologie des Pollens mit besonderer Rücksicht auf die Chemotropische Bewegungen der Polleneschlauche Sitzungsber. *Acad. Wiss. Wien.*, **102**, 428–48.

Panchaksharappa, M.G. & Tirlapur, U.K. (1984). Storage of pollen grains in organic solvents for preserving pollen viability. *J. Palynol.*, **20**, 132–4.

Perry, J.L. & Moore, J.N. (1985). Pollen longevity of blackberry cultivars. *Hort. Sci.*, **20**, 737–8.

Pfahler, P.F. & Linskens, H.F. (1973). *In vitro* germination and pollen tube growth of maize (*Zea mays* L.) pollen. VIII Storage temperature and pollen source effects. *Planta*, **111**, 253–9.

Pfundt, M. (1910). Der Einfluss der Luftfeuchtigkeit auf die' Lebensdauer des Blutenstaubes. *Jahrbuch Wiss. Bot.*, **47**, 1–40.

Rao, A.N. & Chin, L.W. (1972). Experimental studies in the germination of orchid pollen. *Cellule*, **69**, 291–308.

Rao, A.N. & Ong, E.T. (1972). Germination of compound pollen grains. *Grana*, **12**, 113–20.

Rushton, B.S. (1977). Artificial hybridization between *Quercus robur* L. and *Quercus petraea* (Matt.) Liebl. *Watsonia*, **11**, 229–36.

Sedgley, M. (1981). Storage of avocado pollen. *Euphytica*, **30**, 595–9.

Shijun, C. (1984). The study on keeping freshness of orchid pollinia. *Acta Hortic. Sin.*, **11**, 279–80.

Shivanna, K.R. & Heslop-Harrison, J. (1981). Membrane state and pollen viability. *Ann. Bot.*, **47**, 759–70.

Singh, R.K., Tiwara, R. & Kumar, R. (1983). Studies on the pollen viability and storage of drumstick (*Moringa oleifera* Lam.). *Haryana J. Hortic. Sci.*, **12**, 30–5.

Stanley, R.G. & Linskens, H.F. (1974). *Pollen: Biology, Biochemistry and Management*. Berlin: Springer–Verlag.

Vasil, I.K. (1962). Studies on pollen storage of some crop plants. *J. Indian Bot. Soc.*, **41**, 178–96.

Wetzstein, H.Y. & Sparks, D. (1985). Structure and *in vitro* germination of the pollen of pecan. *J. Am. Soc. Hortic. Sci.*, **110**, 778–81.

Wirth, M.W. & Withner, C.L. (1959). Embryology and development in the Orchidaceae. In *The Orchids: A Scientific Survey*, ed. C.L. Withner, pp. 155–88. New York: Ronald Press.

Withers, L.A. (1980). Low temperature storage of plant tissue cultures. *Adv. Biochem. Eng.*, **18**, 101–50.

Effect of temperature and moisture content on the viability of *Cattleya aurantiaca* seed

Introduction

Although epiphytic orchids have been grown routinely from seed for more than sixty years using the asymbiotic method developed by Knudson (1922), relatively little interest has been attached to techniques for the storage of such seed. This is surprising, especially in view of the rapid loss of many orchid habitats, and in particular the loss of tropical moist forest, with the imminent threat of extinction of a large number of orchid species in the wild (Myers 1979, 1980; Koopowitz & Kaye 1983; Hagsater & Stewart 1986; Koopowitz 1986; Stewart 1986). Knudson (1934) indicated the desirability of storing seed to insure against either failure to germinate or the accidental loss of seedlings. The development of such techniques would also allow an assessment of the commercial merits and potential of a particular cross while still retaining a proportion of the seed.

The cryopreservation of seed shows considerable potential. Thus seeds of *Encyclia vitellinum* have been stored at a temperature of –40 °C for 35 days without loss of viability (Koopowitz & Ward 1984). Svihla & Osterman (1943) reported that *Cattleya* hybrid seed survived freezing at –78 °C, and Ito (1965) successfully stored seeds of *Dendrobium nobile* and *Cattleya* hybrids for periods of up to 465 days at –79 °C. Pritchard (1984; 1985) reported that seeds of a number of terrestrial and epiphytic species with seed moisture contents below 14% were not damaged by storage in liquid nitrogen (–196 °C). Although such techniques may be suitable for use in seed banks, it is unlikely that they will be practical for either the amateur or the commercial grower in the foreseeable future. Alternative, more easily applicable techniques for the relatively short term storage of orchid seeds would therefore be of great value.

An increase in mean viability period of many seeds occurs when either seed moisture content, temperature or oxygen pressure of the storage atmosphere have been reduced. Such seeds are described as having orthodox storage characteristics (Roberts 1972). The effects of temperature and seed moisture content in particular have been shown to be interdependent (Owen 1956; Harrington 1960). Thus it was of interest to investigate the effects of temperature and moisture content on the viability of *Cattleya aurantiaca* seed.

Cattleya aurantiaca (Batem. ex Lindl.) P.N. Don (Orchidaceae) was chosen for study because it was known to be cleistogamous and to produce substantial quantities of viable seed (Knudson 1956). Therefore, relatively large amounts of seed were available in which variation among the offspring was kept to a minimum.

Materials and methods

Seed capsules were harvested just prior to dehiscence. Before opening the capsules were swabbed with 70% ethanol, in a laminar flow hood. A sample of the seed was sown immediately, and at this stage surface-sterilization was not necessary. The remainder of the seed was placed in small glass vials, so that each vial contained approximately 400 seeds. The open vials containing seed were placed in small desiccators containing constant-humidity solutions, and the seed allowed to equilibrate for seven days. Saturated solutions of lithium chloride, calcium chloride, sodium dichromate, ammonium chloride and zinc sulphate provided relative humidities of 15, 31, 52, 79 and 90% respectively (Spencer 1926), and seed moisture contents of 3.7, 5.6, 6.5, 10.4 and 14.1% (fresh weight basis) respectively. Equilibration of seed over silica gel produced a seed moisture content of 2.2%. At the end of the equilibration period, the vials were removed from the desiccator, and the air-space in each vial was reduced with a glass sleeper. Each vial was sealed with a small cork bung, which was dipped into molten paraffin wax. Seed samples were stored at either −18 °C, 5 °C or 20 °C.

Procedure used to surface-sterilize and sow seed

All operations were carried out in a laminar flow hood. Seed was placed in a sterile glass tube and surface-sterilized with a solution of 5% (v/v) commercial bleach ('Domestos', Lever Brothers, UK). After 1.5 minutes the seed was poured into a sterile sintered glass crucible, the sterilant removed by suction, and the seed washed with a total of 200 cm^3 sterile deionised water. Seeds were resuspended in 2 cm^3 sterile deionised water, and 0.5 cm^3 aliquots were dispensed using sterile 1 cm^3 disposable syringes. Thus approximately one hundred seeds were transferred to each 125 cm^3 Erlenmeyer flask, and dispersed evenly over the surface of the medium using a swirling motion. The medium employed was that of Thompson (1974). The flasks were plugged with solid rubber bungs and placed in a Warren Shearer controlled environment chamber at a temperature of 22.5 ± 2 °C. Continuous lighting was provided by means of 8 × 80 watt warm white fluorescent tubes and four tungsten filament bulbs, approximately 60 cm above the treatment area, which maintained a fluence rate of about 142 μmol m^{-2} s^{-1}. A random arrangement of flasks was adopted over the treatment area, and the flasks were spaced approximately 10 cm apart.

Measurement of growth

The percentage germination for each treatment was determined 14 and 28 days after sowing by scoring a minimum of 100 seeds in each of four replicate flasks. Seeds were considered to have germinated when the embryos were swollen and green.

Growth was assessed 50 and 100 days after sowing, by measuring protocorm diameter with an eyepiece graticule. At the end of each experiment the degree of development of the seedlings was assessed using an index of protocorm development (Hailes & Seaton 1989; see Chapter 6 this Volume) and both fresh and dry weight of 100 representative seedlings from each flask was determined.

Seed sectioning

The technique used was a modification of that used by Weismeyer & Hofsten (1974). Seed was fixed by submerging in a 4% solution of glutaraldehyde in 0.1M phosphate buffer, pH 7.2, at room temperature for 3 hours, and then washed twice in 0.1M phosphate buffer for 30 minutes per wash. The buffer was replaced with 2% osmium tetroxide, and after two hours the seed was washed twice with 50% propanone. The seed was pelleted by centrifugation, and left in 70% propanone overnight. It was then dehydrated in 90% and 95% propanone for 10 minutes each, and finally three changes of 100% propanone, with centrifugation between each change. The material was then transferred to propylene oxide for 10 minutes, and to fresh propylene oxide for one hour. Following incubation in an equal mixture of propylene oxide and Spurr resin for one hour, the seed was transferred to pure resin at room temperature for three hours, and the resin subsequently cured overnight at 70 °C.

Sections were cut on a Sorvall Porter–Blum MT2–B Ultra-microtome to a thickness of 1.5 μm, and stained with 0.1% toluidine blue. Preparations were examined with a Vickers N41 Photoplan microscope.

Results

Effect of seed moisture content and temperature on viability

Although the two seed samples were obtained from different parent plants, in different years, their survival curves were broadly similar (Figures 1, 2 & 3). At all three storage temperatures, seed stored with a moisture content of 5.6% retained its viability longer than seed stored with a moisture content of 3.7%, which in turn retained its viability longer than seed stored with a moisture content of 2.2%. Indeed Figure 3 shows that seed stored at 5 °C, with moisture contents of 3.7, 5.6, 6.5, 10.4 and 14.1%, retained their initial viability for the whole of the 363 day experimental period. However when stored at 5 °C for periods of six and seven years (Figure 4) seed with a moisture content of 5.6% gave the highest germination. The germination of *C. aurantiaca* seed stored for six years with a 5.6% moisture content was significantly greater than that of 3.7% and 10.4% moisture content seeds (P<0.05), but not significantly greater than seed with a moisture content of 6.5% (P>0.05).

Figures 1, 2 & 3 also show that at moisture contents of 2.2, 3.7 and 5.6%, seed retained its viability longer when stored at 5 °C than at 20 °C; and that seed stored at 20 °C generally retained its viability longer than at –18 °C.

Figure 1. Effect of temperature and moisture content on the viability of *Cattleya aurantiaca* seed stored at –18 °C(A), 20 °C(B) and 5 °C(C), and moisture contents of 2.2(●), 3.7(○), and 5.6% (■). Seed lot 1.

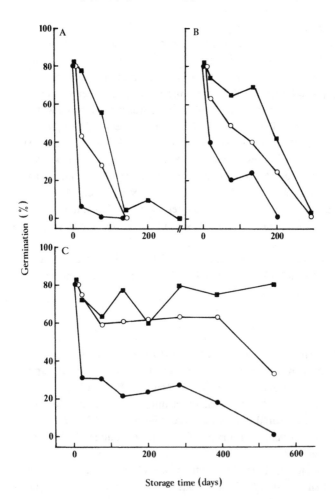

Storage time (days)

Although seed viability was controlled by storage temperature and seed moisture content, if a seed germinated, subsequent seedling growth was unaffected by the storage environment (Tables 1 & 2).

Estimation of number of cells in seed

The cells in the embryo vary in size according to their position (Figure 5). Those of the chalazal or meristematic region are on average 8–10 μm in diameter, whereas the cells of the basal region are larger, and these in turn are attached to a

Figure 2. Effect of temperature and moisture content on the viability of *Cattleya aurantiaca* seed stored at –18 °C(A) and 20 °C(B), and moisture contents of 2.2 (●), 3.7(○), and 5.6% (■). Seed lot 2.

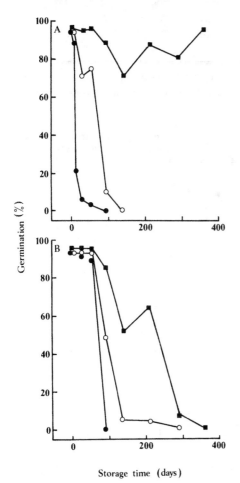

Storage time (days)

Table 1. *Effect of storage temperature and seed moisture content on the development of seedlings measured 120 days after the commencement of germination. Figures in the body of the table represent protocorm development index.[a]*

Temperature (°C)	Moisture content (%)	Storage time[b]				
		0 d	7 d	54 d	363 d	6 y
−18 °C	5.6	214	231	294	235	—
5 °C	2.2	214	251	288	—	—
5 °C	3.7	214	202	287	234	203
5 °C	5.6	214	231	305	243	221
5 °C	6.5	214	204	262	241	195
5 °C	10.4	214	203	247	304	216
5 °C	14.1	214	208	289	278	—

[a]protocorm developmental stages 1–4 multiplied by the number of individuals at each stage (see Hailes & Seaton, Chapter 6 this Volume).
[b]d, days; y, years.

Figure 3. Effect of temperature and moisture content on the survival curves for *Cattleya aurantiaca* seed stored at 5 °C, and 6 moisture contents: 2.2 (●), 3.7 (○), 5.6 (■), 6.5 (□), 10.4(▲) and 14.1% (△). Seed lot 2.

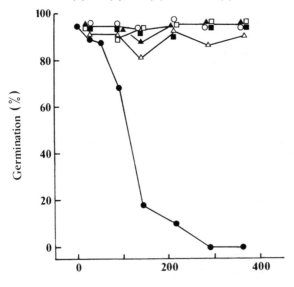

Storage time (days)

Table 2. *Effect of storage temperature and seed moisture content on the development of seedlings measured 120 days after the commencement of germination. Figures in the body of the table represent fresh weight (g) of 400 seedlings.*

Temperature °C	Moisture content (%)	Storage time[a]				
		0 d	7 d	54 d	363 d	6 y
−18 °C	5.6	1.27	1.91	1.97	1.48	—
5 °C	2.2	1.27	2.56	3.17	—	—
5 °C	3.7	1.27	1.49	2.37	1.50	1.80
5 °C	5.6	1.27	1.91	2.49	1.84	2.61
5 °C	6.5	1.27	1.96	1.67	2.18	1.85
5 °C	10.4	1.27	1.16	1.71	1.70	0.86
5 °C	14.1	1.27	2.12	2.92	1.90	—

[a]d, days; y, years.

Figure 4. Effect of seed moisture content on the germination of seed stored at 5 °C for 6 (open column) and 7 years (hatched column).

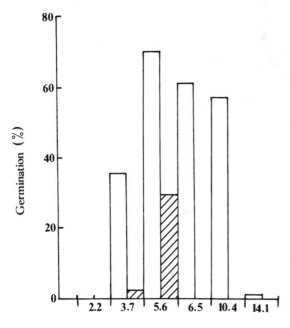

Moisture content (%)

suspensor composed of dead cells (Harrison 1973). The number of cells present in the embryo was calculated by two different techniques. The first method assumed that the embryo consisted of two truncated cones (Figure 6). Each cell was assumed to be a cube, and the area of one face was calculated by dividing the area of one face of each truncated cone by the number of cells it contained. Thus the volume of an average cell could be obtained. The volume of each truncated cone was calculated, and divided by the volume of an average cell, giving a total number of cells per embryo of 123. The second method depended upon obtaining a complete and uninterrupted series of sections in order to trace accurately the appearance and disappearance of individual cells. Because cells were present in more than one section, simply counting the number of cells present in each section would produce an overestimate of the total number of cells present. It was therefore necessary to identify individual cells, and to trace their appearance and disappearance in the serial sections. In order to achieve this, drawings of the section photographs were prepared by tracing the photographs, and reproducing the tracings on plain white sheets (Figure 7). Transparent overlays were

Figure 5. Longitudinal section of *Cattleya aurantiaca* seed.

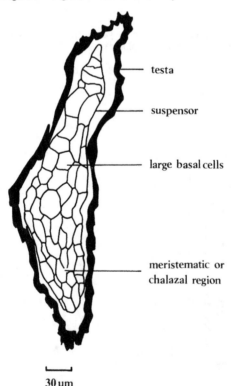

testa

suspensor

large basal cells

meristematic or
chalazal region

30 μm

then prepared from these sheets. Using both sheets and overlays, and by colour-coding each cell as it appeared, it was possible to count all of the cells. This method gave a total of 124 cells in the embryo.

Discussion

Orthodox seeds in general have survival curves which conform to negative cumulative normal distributions i.e. the frequency of individual seed deaths with time is normally distributed (Roberts 1972). In contrast many unicellular resting bodies have survival curves which suggest that each individual is subject to the constant probability of being victim to a lethal accident. Thus the survival curves of spores and non-dividing vegetative cells of bacteria are typically linear when log survivors are plotted against time. The survival curves for colonial bacteria however, are intermediate in form, having a short shoulder and a long tail.

Seeds of *Cattleya aurantiaca* appear to show survival characteristics which are intermediate between those of higher plant seed and those of unicellular bacteria as a number of the curves illustrated in Figure 1, 2 & 3 tend towards the short shoulder and long tail type similar to those observed with colonial bacteria. That the seed

Figure 6. Diagram used to calculate the number of cells in one seed of *Cattleya aurantiaca* showing truncated cones A and B. Testa not shown.

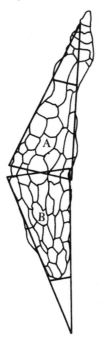

survival curves produced for *Cattleya aurantiaca* did not conform to the negative cumulative normal distribution curves produced by the seeds of other plant families was confirmed by probit analysis (Finney 1971). Furthermore the survival curves were not similar to those of unicellular bacteria, and the data did not produce linear relationships when log survivors was plotted against time.

Relatively few studies have involved an examination of the longevity of orchid seed. However, Humphreys (1960) stated that, in general, orchid seeds lost their viability within nine months. Brummitt (1962) found that *Cypripedium* seed lost its viability within two months, and Lindquist (1965) that the seed of *Disa uniflora* lost

Figure 7. Serial sections A–K, taken transversely through one seed of *Cattleya aurantiaca*. Testa not shown.

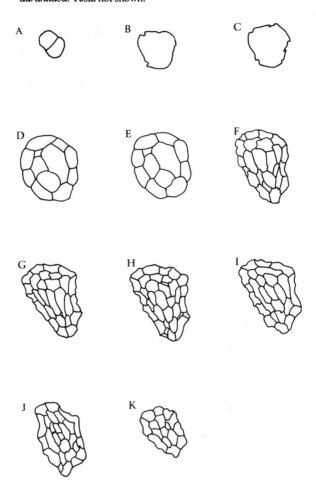

its viability in less than one month. Knudson (1940) reported considerable variation in percentage germination after seed had been stored for between seven and fourteen years. For example, a number of *Brassocattleya* hybrids gave 80% germination after ten years, whereas two *Cattleya* hybrids gave 5 and 6% germination after eight years. Pritchard (1984; 1985) indicated the importance of seed moisture content on seed longevity. It is therefore possible that at least some of the reported variation in longevity of orchid seeds could be due to storage at different moisture contents, and/or storage at different temperatures.

Although *C. aurantiaca seed* was stored successfully at −18 °C for up to one year, this was the least favourable storage temperature at the three seed moisture contents examined. This observation was in contrast to those of other workers, who found that sub-zero temperatures were beneficial for the storage of seed as long as seed moisture contents were not sufficiently high to allow the formation of ice crystals. Thus Ito (1965) reported the successful storage of *Cattleya* and *Laeliocattleya* hybrid seed for more than one year at −79 °C. Bowling & Thompson (1972) stored seeds of thirty different orchid species at −10 °C, reporting that some seed remained viable after three years. However, in 1980 viability tests showed that the seeds were dead (Pritchard 1986). Roberts (1972) considered that the critical maximum moisture content for sub-zero storage of orthodox seeds could be in the region of 15% or slightly above. It is difficult to explain the relatively rapid decline in viability of *Cattleya aurantiaca* seed stored at −18 °C when compared with seed stored at both 5 °C and 20 °C. If freezing damage was responsible for the loss of viability it seems reasonable to expect that the seed would retain its viability longer with decreasing seed moisture content. However seed retained its viability longer at the higher seed moisture contents (Figures 1, 2 & 3).

Storage of seed over silica gel for one week produced a seed moisture content of 2.2%, which brought about a rapid loss of viability. Silica gel is therefore not a suitable dessicant over which to store seeds of *Cattleya aurantiaca* although it is not an uncommon commercial practice. Meeyot & Kamemoto (1969) found that both silica gel and anhydrous calcium chloride reduced the viability of pollen of both *Dendrobium* and *Oncidium*, and they suggested that this may have been due to excessive dehydration. Satisfactory storage of pollen was obtained for up to one year at 45 °F (7 °C) without dehydration. Thus it remains possible that the loss of seed viability at low moisture contents was due to excessive drying. Alternatively, over-rapid rehydration of the seed during the seed sowing procedure may have led to damage. Evidence to support this view comes from the work of Ellis *et al.* (1982) who found that damage to very dry seeds of *Vigna unguiculata* could be largely avoided by equilibration of the seed with atmospheres of 100% relative humidity at 20 °C prior to sowing.

The optimum moisture content for the retention of viability in seed of *C. aurantiaca* was in the region of 5 to 6%. This is within the current range of preservation standards for orthodox seeds (IBPGR 1976). The observation that seed moisture contents of 10.4% and above lead to a more rapid loss of viability is consistent with the observation of Pritchard (1984), who reported that seed of *Eulophia alta* with a moisture content of 23% gave a 20% reduction in germination after two months' storage at 2 °C, whereas seed with a moisture content of 5% showed no loss of viability after six months. Villiers & Edgcumbe (1975) interpreted a similar observation with *Lactuca sativa* in terms of an accumulation of cellular damage at the higher levels of hydration, coupled with an impairment in the efficiency of biochemical repair and turnover processes.

There is considerable evidence to show that physiological changes, chromosomal damage and genetic mutation all occur during storage of orthodox seeds; but genetic changes may not become evident until the second generation (Roberts 1981). Figures 1, 2 & 3 show that some of the seed storage conditions clearly reduced the percentage of seeds germinating. However, if a seed germinated there was no evidence that the seedling showed reduced vigour during the first 120 days of growth (Tables 1 and 2). This effect was observed not only when growth was monitored using a development index, but also when dry and fresh weight accumulation was determined. In contrast, Pritchard (1985) found an extension of the time to 50% germination in *Eulophia alta* seed after two months' storage under unfavourable conditions.

Acknowledgements
Thanks to Prof. E.H. Roberts for his invaluable comments and advice.

References
Bowling, J.C. & Thompson, P.A. (1972). On storing orchid seed. *The Orchid Rev.*, **80**, 120–1.

Brummitt, L.W. (1962). Cypripediums. *The Orchid Rev.*, **70**, 181–3, 249–51, 384–7.

Ellis, R.H., Osei-Bonsu, K. & Roberts, E.H. (1982). Desiccation and germination of seed of cowpea (*Vigna unguiculata*). *Seed Sci. Technol.*, **10**, 509–15.

Finney, D.J. (1971). *Probit analysis*. Cambridge: Cambridge University Press.

Hagsater, E. & Stewart, J. (1986). Orchid Conservation at the International Level, Part 3. *The Orchid Rev.*, **94**, 123–5.

Hailes, N.S.J. & Seaton, P.T. (1989). The effects of the composition of the atmosphere on the growth of seedlings of *Cattleya aurantiaca*. In *Modern Methods in Orchid Conservation. The Role of Physiology, Ecology and Management*, ed. H.W. Pritchard, pp. 73–85. Cambridge: Cambridge University Press.

Harrington, J.F. (1960). Drying, storing, and packaging seeds to maintain germination and vigor. In *Proc. Short Course Seedsmen*, pp. 89–108. Mississippi: State College.

Harrison, C.R. (1973). Physiology and ultrastructure of *Cattleya aurantiaca* (Orchidaceae) Germination. Ph.D. Dissertation, University of California, Irvine.

Humphreys, J.L. (1960). Help wanted. *The Orchid Rev.*, **68**, 141–2.

International Board for Plant Genetic Resources (1976). *Report of IBPGR Working Group on Engineering, Design and Cost Aspects of Long-term Seed Storage Facilites.* Rome: IBPGR.

Ito, I. (1965). Ultra-low temperature storage of pollinia and seeds of orchids. *Japan Orchid Soc. Bull.*, **11**, 4–15.

Knudson, L. (1922). Nonsymbiotic germination of orchid seeds. *Bot. Gaz.*, **73**, 1–25.

Knudson, L. (1934). Storage and viability of orchid seed. *Am. Orchid Soc. Bull.*, **2**, 66.

Knudson, L. (1940). Viability of orchid seed. *Am. Orchid Soc. Bull.*, **9**, 36–8.

Knudson, L. (1956). Self-pollination in *Cattleya aurantiaca* (Batem.) P.N. Don. *Am. Orchid Soc. Bull.*, **25**, 528–32.

Koopowitz, H. (1986). A gene bank to conserve orchids. *Am. Orchid Soc. Bull.*, **55**, 247–50.

Koopowitz, H. & Kaye, H. (1983). *Plant Extinction: A Global Crisis.* Washington: Stone Wall Press.

Koopowitz, H. & Ward, R. (1984). A technological solution for the practical conservation of orchid species. *Orchid Advocate*, **10**, 43–5.

Lindquist, B. (1965). The raising of *Disa uniflora* seedlings in Gothenburg. *Am. Orchid Soc. Bull.*, **34**, 317–9.

Meeyot, W. & Kamemoto, H. (1969). Studies on storage of orchid pollen. *Am. Orchid Soc. Bull.*, **38**, 388–93.

Myers, N. (1979). *The Sinking Ark.* Oxford: Pergamon Press.

Myers, N. (1980). *The Conservation of Tropical Moist Forests.* Washington D.C.: National Academy of Sciences.

Owen, E.B. (1956). The storage of seeds for maintenance of viability. *Commonw. Bur. Pastures Field Crops Bull.*, **43**, 1–81.

Pritchard, H.W. (1984). Liquid nitrogen preservation of terrestrial and epiphytic orchid seed. *Cryo-Lett.*, **5**, 295–300.

Pritchard, H.W. (1985). Growth and storage of orchid seeds. *Proc. 11th World Orchid Conference, Miami, 1984*, ed. K.W. Tan, pp. 290–3. 11th World Orchid Conference, Inc.

Pritchard, H.W. (1986). Orchid seed storage at the Royal Botanic Gardens, Kew, England. 2. Physiology Unit, Wakehurst Place. *Orchid Res. Newsl.*, **7**, 18.

Roberts, E.H. (1972). Storage environment and the control of viability. In *Viability of Seeds*, ed. E.H. Roberts, pp. 14–58. London: Chapman and Hall.

Roberts, E.H. (1981). Physiology of ageing and its application to drying and storage. *Seed Sci. Technol.*, **9**, 359–72.

Spencer, H.M. (1926). Laboratory methods for maintaining constant humidity. In *International Critical Tables 1*, pp. 67–8. New York: McGraw–Hill Book Co.

Stewart, J. (1986). Orchid conservation at the international level. *Am. Orchid Soc. Bull.*, **55**, 242–6.

Svihla, R.D. & Osterman, E. (1943). Growth of orchid seeds after dehydration from the frozen state. *Science*, **98**, 23–4.

Thompson, P.A. (1974). Orchids from seed, a new basal medium. *The Orchid Rev.*, **82**, 179–83.

Villiers, T.A. & Edgcumbe, D.J. (1975). On the cause of seed deterioration in dry storage. *Seed Sci. Technol.*, **3**, 761–74.

Weismeyer, H. & Hofsten, A.V. (1974). Electron microscopy of orchid seedlings. In *First Symposium on the Scientific Aspects of Orchids*, ed. H.H. Szmant & J. Wemple, pp. 16–26. Southfield, Michigan: University of Detroit.

Asymbiotic germination of epiphytic and terrestrial orchids

Introduction

The application of *in vitro* techniques to the propagation of plants at Kew originated from Dr Peter Thompson's work in the 1960s on the formulation of a medium for orchid seed germination. By 1971 the commercial and research potential of plant propagation by aseptic culture was becoming more apparent and the Director of Kew at the time, Professor Heslop-Harrison, determined that it was time to make the techniques available to the living collection. Plans were made for a unit to provide *in vitro* propagation services to the sections of the Living Collections Division at Kew. The Unit based at Aiton House was set up and opened in 1977, under the present Curator, Mr John Simmons.

The living orchid collection at Kew comprises approximately 3,500 species represented in 375 genera. Each year the resultant seed from a pollination programme is germinated at the Micropropagation Unit. Seed is also given to the Unit by botanists from field collections, from private and commercial orchid growers and from members of the general public. The main function of the Unit is to supply the orchid collection with new species and in many cases provide a back-up of viable seedlings for those species which present special cultivation problems. As much information as possible is recorded about the seed including such details as the site and habitat type where the collection was made. This type of information helps decide the type of culture conditions used. Natural source seed is of particular importance, since it has considerable genetic potential. Collections from native habitats are always made with the full agreement and co-operation of the relevant authorities and in many cases Kew agrees to return some of the seedlings raised to the national parks from which they originally came.

Techniques

Seed collecting, storage and treatment

Material is usually received as dehisced or undehisced capsules, or as loose seed. Seed collected in the field may become contaminated by fungal infection during transit. As this is most likely to occur under moist, warm conditions, it is most important to dry the seed and capsules as soon after harvest as possible. Temporary storage of seed is best in small paper bags or unsealed vials; storage in plastic bags or sealed vials inevitably traps moisture and creates ideal conditions for fungal spore germination and seed infection. Undehisced or loose seed is usually sieved to remove any debris, and then placed into pre-sterilized vials with cotton wool bungs. Seeds are stored in a desiccator at room temperature until required for sowing. This method of storage seems to be adequate for short periods of several months. However, if seed needs to be stored for longer periods, it is transferred to a cold store at 4°C. It is important to reduce the moisture content of seed prior to storage not only to impede fungal spore germination and growth, but also to enhance seed longevity.

A useful technique for sowing small quantities of rare seed is to sterilize the seed in small filter paper packets. These are immersed for 10 minutes in 3% sodium hypochlorite solution plus wetting agent, and then rinsed in sterile deionised water. Seed can then be spread over the surface of the medium or transferred to the medium on the paper. This method cuts down on seed wastage.

It is usual to sow seed from intact capsules directly on receipt as the surface-sterilization procedure is simpler (5% sodium hypochlorite for 5 minutes) and contamination of seed is usually reduced. It has been reported (Fast 1982) that inhibitory factors which appear to be present in the mature seed coat of some species are absent or ineffective in seed from immature capsules, as in many cases improved germination has been obtained. Improved germination has also been observed in some species, following surface-sterilization of the seeds. For example, in some *Paphiopedilum* species this treatment appears to increase germination by improving the permeability of the testa (Haas-von Schmude *et al.* 1986). Pre-soaking in sterile deionised water also improves germination in *Cypripedium* (Fast 1982). This may be due to inhibitory factors being leached from the seed coat. Embryo dissection has proved to be effective in aiding germination in *Cypripedium calceolus* L. Following sterilization, seed coats are removed using the fine cutting edge of a hypodermic syringe and the seeds soaked in sterile deionised water for 30 to 58 days. Germination has been obtained on both Curtis (1936) and Norstog (1973) medium after approximately 5 months at 22 °C in total darkness.

Medium composition

Many species can be germinated using conventional asymbiotic methods. With such a wide range of species grown at Kew, it is difficult to tailor a medium for each individual species. To a large extent we use commercial formulations and add our own supplements as required. Various media are used. The most successful include Vacin & Went (1949), Curtis (1936), Fast (1976), Murashige & Skoog (1962), Knudson C (1946), Mead & Bulard (1979), Norstog (1973) and Hills' seed sowing and replating media (commercial brands) (Table 1). There have been varying degrees of success with this media. From general observations, the best germination results

Table 1. *Summary of media used at Kew.*

Medium	Germination	Growing-on
Vacin & Went (1949):	+	+
modified with banana		
or coconut milk	—	+
Curtis (1936)	+	—
Fast (1976)	+	—
Murashige & Skoog (1962)	—	+
Knudson C (1946)	+	—
Mead & Bulard (1979)	+	+
Norstog (1973)	+	—
Hills' seed sowing medium[a]	+	—
Hills' replating medium[a]		
(commercial brands)	—	+

[a]available from Daniel M. Hill, P.O. Box 1184, Ontario, California 91762, USA.

have been achieved on Vacin & Went, Fast and Norstog. For growing on, Hills' replating medium is good for *Paphiopedilum*, possibly stimulated by the presence of biotin in the medium (Lucke 1971).

The addition of homogenized banana and/or coconut milk to Vacin & Went salts has been beneficial to growth in many species, encouraging strong shoots and roots. However, with *Dactylorhiza fuchsii* (Druce) Soó the addition of banana to the medium causes stunted shoot growth. Analysis of coconut milk (Dix & van Staden 1982) and banana (Arditti 1968) has indicated that the active constituents may be auxin and gibberellin-like substances.

Addition of vegetable charcoal to medium at concentrations between 2 and 5 g dm^{-3} has been indicated to improve the growth of many genera, including *Paphiopedilum* (Ernst 1974). From general observations on several species of *Paphiopedilum*, we have observed that the addition of charcoal results in chlorotic growth. In *Phalaenopsis* cultures charcoal appears to absorb the phenolic compounds exuded by the roots (Ernst 1976). Charcoal may also improve aeration of the medium (Arditti & Ernst 1984).

Carbohydrate is normally supplied as sucrose, fructose or glucose, although others used include mannose, raffinose, mannitol and trehalose. The latter two are particularly efficient as they are of fungal origin (Smith 1973). The former are usually included in media at concentrations between 5 and 30 g dm^{-3}. It is probable that the relative concentrations of these constituents alters during autoclaving, as sucrose is known to be hydrolyzed on autoclaving to give a combination of sucrose, fructose and glucose (Helgeson *et al.* 1972). This may well affect the germination achieved.

Similarly, changes in pH have been shown to occur during autoclaving of media; this needs to be taken into account where pH is of particular importance (Skirvin *et al.* 1986).

Culture vessels and environmental conditions

Various culture vessels have been tried at Kew, including Petri dishes, conical flasks, Kilner jars and screw top jars. The type of vessel used in culture should be chosen carefully as this will influence the internal humidity of the system (Debergh 1983; Wardle *et al.* 1983). This may further influence cuticle development in the seedlings (Sutter 1985) and also the period for which the culture can be maintained. For sowing, pre-sterilized Petri dishes are useful; they make handling more efficient and examination under a binocular microscope easier. Dishes with moulded compartments are useful when making comparisons between media. All dishes should always be well sealed to prevent drying out of the medium.

Because of the wide range of species grown at Kew, general culture conditions are employed, viz. 22–25 °C with a 16 hour photoperiod. Light does not appear to have an inhibitory effect on germination in epiphytes (Arditti 1979), however it has been reported to inhibit germination in a number of terrestrial species, including *Cypripedium* and *Orchis* spp. (Harvais 1973; Fast 1976). These species are usually germinated in total darkness.

Germination and seedling morphology

Detailed studies on germination and morphology in seedlings of *Cymbidium ensifolium* (Tian *et al.* 1985) have highlighted the need for standardized use of nomenclature when describing the morphology of orchid seedlings. Moreover, it would be useful to adopt a standard growth index for developing seedlings, perhaps along the lines of those used by Hailes & Seaton, Chapter 6 this Volume. Such a system would enhance reports such as the recent study on *Paphiopedilum rothschildianum* Reichb. f. (Haas-von Schmude *et al.* 1986) which produced some interesting observations on culture and development of seedlings, but lacked this type of detail. Additionally, these details might allow interesting comparisons to be made with other reports on the successful propagation of the same species (e.g. for *P. rothschildianum*, Ernst 1974).

Species case histories

The production of plants from seed has advantages to the grower and conservationist. It is the best way of conserving natural populations whilst still increasing genetic diversity and producing strong healthy plants (see Figure 1). Over the past few years examples of this method of propagation for rare or threatened orchids can be found from both botanic gardens and commercial establishments. The first examples given here are of seedlings raised at Kew.

Orchis laxiflora *Lam.*

This European terrestrial orchid, now confined to southern Europe and parts of France once occurred as far north as the Channel Islands. Germination was achieved on both Mead & Bulard medium and Vacin & Went medium at 22 °C with a 16 hour photoperiod. After 12 weeks seedlings were transferred to larger vessels with the same medium. After a further 8 weeks seedlings had developed good roots and were ready to transfer to the glasshouse. They were potted into a fine bark, charcoal, peat and terra-green mix. After an initial period of 2 weeks in a dew-point cabinet at 90% humidity and with supplementary lighting, seedlings established readily in a cool glasshouse at 11 °C.

Cypripedium calceolus *L.*

This elegant species, although fairly widespread throughout Europe, is now

Figure 1. Flow diagram showing the advantages of growing orchids from seed.

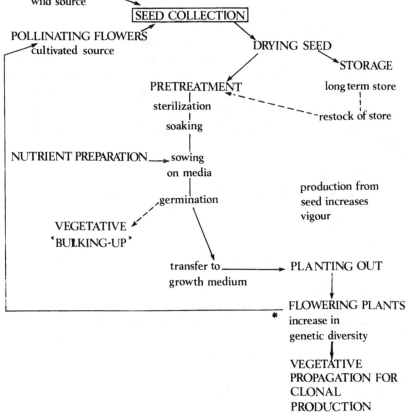

reduced to one remaining site in the U.K. Gradual loss of habitat through intensive agricultural methods, and collection of the species for cultivation has resulted in this current state. Germination has been achieved in this species by several workers (Fast 1976; Lucke 1976; Butcher unpublished). However the germination seems to be slow and problematic.

The reintroduction of this species is an exciting goal to aim for. However a great deal of work still needs to be done before this aim can be realised.

Eulophia toyashimae *Nakai*.

This unusual subterranean saprophyte is only recognizable when the flower spike is sent above the ground. Seed from the last reported flowering plant sighting was germinated on Vacin & Went medium at 22 °C, transferred and grown on. In such cultures the rhizomatous growth habit can be seen. There is a planned reintroduction program instigated by Keith Wooliamson of the Waimea Arboretum, Hawaii. Under the storage conditions described earlier seed has remained viable for more than three years.

Dendrobium spectatissimum *Reichb.f.*

Seed of this rare species was collected in Borneo by Dr Phillip Cribb, and has been successfully germinated on Vacin & Went medium. Seedlings were transferred to modified Vacin & Went medium for a further 16 weeks before they were ready to pot in a fine bark, perlite and charcoal compost. Seedlings have now been returned to the National Park Authorities in Sabah, and it is hoped that surplus seedlings will be available for distribution to other establishments in the near future.

Cattleya dowiana *var.* aurea *Batem.*

This high conservation rated species from South America, well known amongst growers as a successful parent of many fine quality hybrids, was germinated on Vacin & Went medium and transferred to Vacin & Went with 60 g dm^{-3} banana. Seedlings were then potted into a bark, charcoal and vermiculite compost 12 months after germination. Many plants have now been sent to other botanic gardens.

Paphiopedilum sukhakulii *Schoser & Senghas*.

Seed donated to Kew by Pandora Sellers, germinated on both Fast and Vacin & Went media. After approximately three months, seedlings were transferred to Hills' replating medium or modified Vacin & Went. After a further nine months' growth, seedlings were potted into a fine bark, loam and fine gravel compost.

A summary of the germination and growing-on media used at Kew is presented in Table 1.

The next example given is from a commercial establishment. Similar examples are to be found within many of the well known orchid nurseries.

Paphiopedilum rothschildianum *Reichb f.*

This species is extremely rare and threatened in its native habitat, on Mount Kinabalu, Borneo. The plant grown at Ratcliffe Orchids Ltd, was believed to be a division of a plant collected by Sander in Borneo. In 1973 the plant was successfully selfed and approximately 50 seedlings were produced. Dr. J. Binks, an established amateur grower, purchased and grew on 25 of these seedlings. The plants began to flower in 1981 and in 1985 one of the plants was awarded an Award of Merit by the Royal Horticultural Society. Following this the plant was crossed with one of its siblings resulting in excellent seed production. The seed was germinated on Hills' seed sowing medium and grown-on on Hills' replating medium. The seedlings are very vigorous growers, probably a result of the cross pollination. It is estimated that approximately 2000 seedlings have now been sold worldwide at very reasonable prices. There is now no reason why anyone should need to buy a wild source plant of this species.

Concluding remarks

Many orchid species are threatened by the indiscriminate collection which continues, despite the high conservation rating they now hold using the scale by the International Union for the Conservation of Nature and Natural Resources (IUCN). For example, *Paphiopedilum micranthum*, re-discovered as recently as 1980, is now reported to be extinct in its native habitat. Perhaps it is too late for the protection of many of these beautiful species. However, if the plants already held in collections could be used to produce seed which could be made available to botanic gardens and commercial growers alike, through national and international seed banks, then, hopefully, some protection may be afforded to the remaining few stands of these valuable plants.

References

Arditti, J. (1968). Germination and growth of orchids on banana fruit tissue and some of its extracts. *Am. Orchid Soc. Bull.*, **37**, 112–6.

Arditti, J. (1979). Aspects of the physiology of orchids. *Adv. Bot. Res.*, **7**, 421–655.

Arditti, J. & Ernst, R. (1984). Physiology of germinating orchid seeds. In *Orchid Biology Reviews and Perspectives III*, ed. J. Arditti, pp. 197–200. Ithaca, USA: Cornell University Press.

Curtis, J.T. (1936). The germination of native orchid seeds. *Am. Orchid Soc. Bull.*, **5**, 42–7.

Debergh, P.C. (1983). Effects of agar brand and concentration on the tissue culture medium. *Physiol. Plant.*, **59**, 270–6.

Dix, L. & van Staden, J. (1982). Auxin and gibberellin-like substances in coconut milk and malt extract. *Plant Cell Tissue and Organ Culture*, **1**, 239–46.

Ernst, R. (1974). The use of activated charcoal in asymbiotic seedling culture of *Paphiopedilum*. *Am. Orchid Soc. Bull.*, **43**, 35–8.

Ernst, R. (1976). Charcoal or glass wool in asymbiotic culture of orchids. In *Proc. 8th World Orchid Conf.*, ed. K Senghas pp. 379–83. Frankfurt am Main: German Orchid Society Inc.

Fast, G. (1976). Moglichkeiten zur Massenvermehrung von *Cypripedium calceolus* und anderen europäischen Wild orchideen. In *Proc. 8th World Orchid Conf.*, ed. K. Senghas, pp. 359–63. Frankfurt am Main: German Orchid Society Inc.

Fast, G. (1982) European terrestrial orchids (symbiotic and asymbiotic methods). In *Orchid Biology Reviews and Perspectives II*, ed. J. Arditti, pp. 309–26. Ithaca, USA: Cornell University Press.

Haas-von Schmude, N.F., Lucke, E., Ernst, R. & Arditti, J. (1986). *Paphiopedilum rothschildianum. Am. Orchid Soc. Bull.*, **55**, 579–84.

Harvais, G. (1973). Growth requirements and development of *Cypripedium reginae* in axenic cultures. *Can. J. Bot.*, **51**, 327–32.

Helgeson, J.P., Upper, C.D., & Haberlach, G.T. (1972). Medium and tissue sugar concentrations during cytokinin-controlled growth of tobacco callus tissues. In: *Plant Growth Substances 1970*, ed. D.J. Carr, pp. 484–92. Berlin: Springer–Verlag.

Knudson, L. (1946). A new nutrient solution for the germination of orchid seed. *Am. Orchid Soc. Bull.*, **15**, 214–7.

Lucke, E. (1971). The effect of biotin on sowings of *Paphiopedilum. Am. Orchid Soc. Bull.*, **40**, 24–6.

Lucke, E. (1976). Erste Ergebnisse zur asymbiotischen Samenkeimung von *Himantoglossum hircinum. Die Orchidee*, **27**, 60–1.

Mead, J.W. & Bulard, C. (1979). Vitamins and nitrogen requirements of *Orchis laxiflora* Lamk. *New Phytol.*, **83**, 129–36.

Murashige, T. & Skoog, F. (1962). A revised medium for rapid growth and bio assays with tobacco tissue cultures. *Physiol. Plant.*, **15**, 473–97.

Norstog, K. (1973). New synthetic medium for the culture of premature barley embryos. *In Vitro*, **8**, 307–8.

Skirvin, R.M., Chu, M.C., Mann, M.L., Young, H., Sullivan, J. & Fermanian, T. (1986). Stability of tissue culture medium pH as a function of autoclaving, time, and cultured plant material. *Plant Cell Rep.*, **5**, 292–4.

Smith, S.E. (1973). Asymbiotic germination of orchid seeds on carbohydrates of fungal origin. *New Phytol.*, **72**, 497–9.

Sutter, E.G. (1985). Morphological, physical and chemical characteristics of epicuticular wax on ornamental plants regenerated *in vitro. Ann. Bot.*, **55**, 321–9.

Tian, M.S., Wang, F.X., Qian, N.F. & Sun, A.C. (1985). *In vitro* germination and developmental morphology of seedlings in *Cymbidium ensifolium. Acta Bot. Sin.*, **27**, 455–9.

Vacin, E.F. & Went, F.W. (1949). Some pH changes in nutrient solutions. *Bot. Gaz.*, **110**, 605–13.

Wardle, K., Dobbs, E.B. & Short, K.C. (1983). *In vitro* acclimatization of aseptically cultured plantlets to humidity. *J. Am. Soc. Hortic. Sci.*, **108**, 386–9.

Germination and mycorrhizal fungus compatibility in European orchids

Introduction

The project at Kew is concerned primarily with the symbiotic method of raising European orchids from seed. However, asymbiotic sowings have occasionally been made for direct comparison of the relative effectiveness of the two methods, and in an attempt to raise seedlings where the symbiotic method has proved unsuccessful. This paper describes a comparison between asymbiotic and symbiotic germination of three species of *Orchis*, for which both methods were successful. In addition, the effective asymbiotic methods for germination of rare British species, where symbiotic methods have failed, and the germination response of three orchid species from each of the genera *Orchis*, *Ophrys*, *Dactylorhiza* and *Serapias* to nine vigorous and eight less vigorous orchid symbionts are reported. The pattern of orchid/fungus compatibility is also discussed in relation to the raising of seedlings beyond initial germination stages, and the routinely used orchid/fungus combinations for bulk propagation of certain species are recorded and illustrated.

Materials and methods

All sowings were made on to agar-based media in Petri dishes using aseptic techniques (Muir 1987). For symbiotic germination the media used were Modified Oats Medium (O3) (Clements *et al.* 1986) and G4 – a modification of O3, using 1.2 g dm^{-3} amylopectin in place of the oats, on the recommendation of P. Milon (Laboratoire de Recherches Horticoles, 78570 Chanteloup-les-Vignes, Paris). The media of Harvais (1973), Mead & Bulard (1975), Norstog (1973) and Curtis (1936) were used for asymbiotic germination. 200–600 seeds were sown per treatment in 3–6 replicate plates. All were incubated in the dark at 21 ± 1 °C.

All mycorrhizal fungi used were isolated at Kew from the root tissue of mature orchids in cultivation, with the exception of cultures Cc – a strain of *Ceratobasidium cornigerum* obtained from the Commonwealth Mycological Institute, and M10 – an *Ophrys* symbiont donated by P. Milon. All Kew isolated fungi were assigned Q numbers.

Germination was assessed on a scale of 0–5, as previously described (Clements *et al.* 1986), and illustrated in Figure 1.

Propagation beyond the initial stages of germination was continued *in vitro* in either Geneco vessels or Bunzl flasks, as described previously for *Orchis laxiflora* Lam (Muir 1987).

Results and discussion

Germination of Orchis *species*
Three species of *Orchis* were sown on three asymbiotic media, and on two

Figure 1. Stages used to assess germination 0: ungerminated seed. 1: germination of seed with rupture of testa. 2: production of rhizoids. 3: production of leaf primordium. 4: production of first chlorophyllous tissue. 5: production of root initial. 0.5 mm scale bar for stages 0 and 1, 1 mm scale bar for stages 2–5.

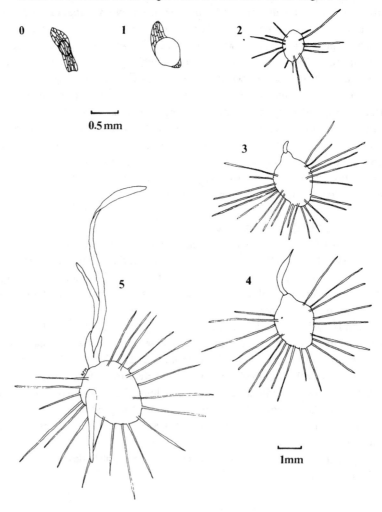

Table 1. *Total asymbiotic and symbiotic germination (%) of* Orchis *spp. after 5 weeks' incubation in the dark at 21 °C.*

Medium		*Orchis laxiflora*	*Orchis morio*	*Orchis sancta*
Asymbiotic:				
Mead & Bulard (1975)		62.5	86.5	54.6
Norstog (1973)		42.7	79.7	84.1
Harvais (1973)		49.4	68.3	51.9
Symbiotic:	Fungus			
O3[a]	Q14	51.7	74.0	73.6
G4[b]	"	36.2	77.8	70.5
O3	Q21	41.2	40.3	79.8
G4	"	37.1	67.0	84.2
O3	Q23	43.3	78.5	70.7
G4	"	34.1	50.6	80.4
O3	Cc	26.7	47.6	48.4
G4	"	28.3	63.5	50.0

[a] O3 = Modified Oats Medium (Clements *et al.* 1986)
[b] See materials and methods for details

symbiotic media with four different mycorrhizal fungi. The species chosen: *Orchis laxiflora, Orchis morio* L. and *Orchis sancta* L. had all been found in preliminary trials to be fairly fast-developing *in vitro* and five weeks was considered to be a suitable time at which to assess development, when a range of protocorm stages would be likely to be found in any one sowing. Three very different asymbiotic media, commonly used for European orchid germination, were chosen for comparison: Norstog, high in amino acids; Harvais, containing potato extract as a complex additive; and Mead & Bulard, containing casein hydrolysate as a complex additive, and specially formulated for *Orchis laxiflora*. The four fungi chosen were ones previously found to be compatible and derived from different source species.

Table 1 shows the total germination for *Orchis laxiflora, Orchis morio* and *Orchis sancta* for each treatment. The viability of *Orchis laxiflora* appears to be lower than that of the other two species, as germination is lower in all treatments. Overall, asymbiotic and symbiotic methods with these three species give similar results, though Mead & Bulard is superior for the germination of *Orchis laxiflora* and *Orchis morio*. The best symbiotic combinations are: O3 and Q14 for *Orchis laxiflora*, O3 and Q23 for *Orchis morio*, and G4 and Q21 for *Orchis sancta*.

However, when the stages of germination reached are taken into account, a slightly different picture emerges. Figure 2 shows the development of each of the species on asymbiotic media. For *Orchis laxiflora* (Figure 2A), the furthest stage of development

Figure 2. Stages of development of *Orchis laxiflora* (A), *O. morio* (B) and *O. sancta* (C) five weeks after sowing seed on Mead & Bulard (□), Norstog (▨) and Harvais (▩) medium. Germination conditions: 21 °C in the dark.

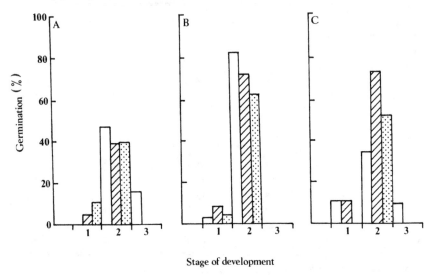

Figure 3. Symbiotic germination (%) of *Orchis laxiflora* (A), *O. morio* (B), and *O. sancta* (C) on O3 (▨) or G4 (□) medium with fungal isolates Q14, Q21 and Q23 isolated at Kew, and Cc (*Ceratobasidium cornigerum*). Germination details as Figure 2.

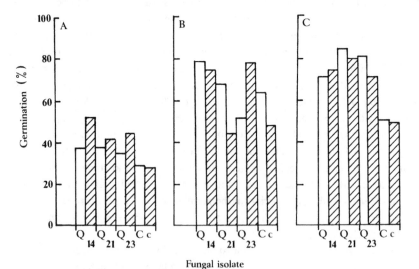

is found with Mead & Bulard medium with all protocorms at either stage 2 or 3 and none remaining at 1, whereas with the other two media the protocorms only reached stages 1 or 2. For *Orchis morio* (Figure 2B), results on the three media are similar, with Mead & Bulard slightly more effective. For *Orchis sancta* (Figure 2C), although the best medium for total germination appeared to be Norstog, the most advanced stage is found on Mead & Bulard. It is important to recognise these differences in degree of development, over the given time period, as it is the most advanced protocorms which will eventually go on to produce mature seedlings, and this may not be always on the medium which initially gives the highest germination. Often media will support germination up to stage 2, but no further. Therefore, for any germination experiment designed to determine the most successful media for raising seedlings to maturity, development must be assessed beyond stage 2 and preferably to stage 4.

Figure 3 shows *Orchis* species germination on the symbiotic media, with each of the isolates. There is very little difference between the total germination response on the two media, though perhaps O3 gives slightly higher figures with *Orchis laxiflora* and fungi Q14 and Q23 (Figure 3A), and G4 gives slightly higher figures with *Orchis sancta* and fungus Q23, though for both species Cc seems less effective. With *Orchis morio* results are more affected by media differences. The most effective symbionts are Q14 and Q23 on O3, and Q14 and Q21 on G4 (Figure 3B).

If the stages of germination reached are taken into account, as for the asymbiotic media, then differences between the symbiotic sowing become more apparent. Figure 4 shows the stages of development of *Orchis laxiflora* in the symbiotic sowings. On O3, Q21 and Q23 stimulate the development to a greater extent than Q14, with most protocorms at stage 3, and many at stage 4. Also, the least effective isolate for overall germination (Cc, Figure 2) gives the highest number of stage 4 protocorms. On G4,

Figure 4. Stages of development of *Orchis laxiflora* sown symbiotically on O3(A) or G4(B) medium with fungal isolates Q14 (▢), Q21 (▨), Q23 (▨) and Cc (▨). Germination details as Figure 2.

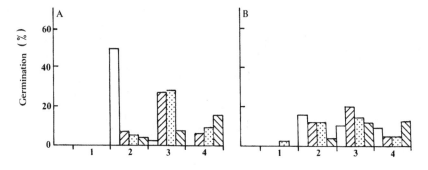

Stage of development

Figure 5. Stages of development of *Orchis morio* sown symbiotically. Details as Figure 4.

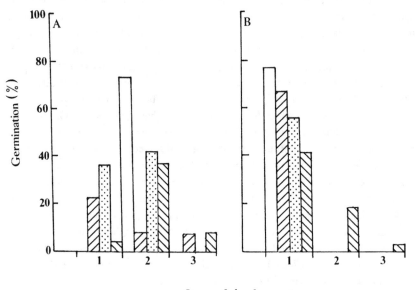

Stage of development

Figure 6. Stages of development of *Orchis sancta* sown symbiotically. Details as Figure 4.

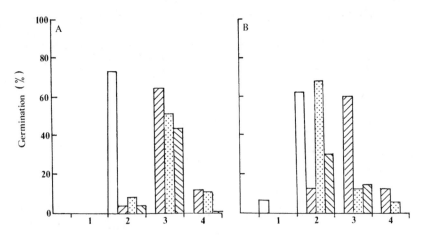

Stage of development

Q14 appears more effective as more protocorms reached stage 4 than with Q21 and Q23, but Cc again produced the highest number of stage 4 protocorms.

Figure 5 shows the stages of development of *Orchis morio* in the symbiotic sowings. Although three of the fungi (Q14, Q21 and Q23) appeared to be more effective for total germination (including all stages) on G4, this medium actually produced a less rapid development than O3 with protocorms only reaching stage 1. Cc stimulated *Orchis morio* to develop most quickly on both media, and to reach the furthest stages, despite the fact that total germination was lower with this isolate than with the others. On O3, Q21 was as effective in producing stage 3 protocorms as Cc.

Figure 6 shows the stages of development of *Orchis sancta* in the symbiotic sowings. As with *Orchis laxiflora*, the least effective symbiont for total germination was Cc, though in contrast it proved less effective in producing stage 3 and 4 protocorms. On both media, Q21 and Q23 resulted in the furthest stages of protocorm development, and these isolates were also the ones which gave the highest germination overall. Though a high overall germination was achieved with Q14, it appeared to be relatively unsuitable for the production of mature protocorms.

Where successful symbiotic methods have been established they compare very favourably with the alternative asymbiotic methods.

Rare British species

In the case of the rare British species *Orchis militaris*, L. the symbiotic method is the only one which has so far proved successful. The stages reached are shown in Figure 7A. However, there are difficulties in discovering effective symbionts, even though large numbers of isolates may be screened (there are over 120 in the Kew collection). Moreover, the rarer the species of orchid, the more exacting it appears to be for both media components and for mycorrhizal fungi.

With the two rare British species, *Cypripedium calceolus* L. and *Himantoglossum hircinum* (L.) Sprengel, symbiotic methods of raising seedlings have not yet been established, but germination has been achieved on asymbiotic media. With *Cypripedium calceolus* effective and reproducible methods have been developed using Harvais, and Curtis media, modified with the addition of 100 g dm^{-3} coconut water. A batch of protocorms seven months from sowing is shown in Figure 7D. *Himantoglossum hircinum* responded to both Norstog and Mead & Bulard media. The stages of development reached are shown in Figure 7B, with some 12 month old seedlings sufficiently advanced for transfer to soil mix *in vitro*.

With the rare British species *Ophrys sphegodes* Miller, both symbiotic and asymbiotic methods have proved successful and reproducible, though so far the asymbiotic method with Mead & Bulard medium has been the most reliable for high levels of germination (55–60%) and speed of development. Figure 7C shows three month protocorms on Mead & Bulard medium immediately after transfer from the sowing vessel to a flask.

Figure 7. Development of symbiotic cultures of *Orchis militaris* (A), and asymbiotic cultures of *Himantoglossum hircinum* (B), *Ophrys sphegodes* (C) and *Cypripedium calceolus* (D). Time from sowing: A, 9 months; B, 12 months, with more advanced seedlings transferred to soil mix *in vitro*; C, 3 months; D, 7 months. Scale bars = 2 cm.

Table 2. *Source of mycorrhizal fungi used in compatibility testing with orchids.*

Mycorrhizal fungus isolate	Orchid species source of isolate
Q3	*Ophrys holoserica* (Burm. f.) Greuter
Q6	*Dactylorhiza romana* (Sebast.) Soó
Q8	*Orchis militaris*
Q13	*Orchis papilionacea* L.
Q14	*Orchis papilionacea*
Q17	*Orchis simia* Lam.
Q18	*Serapias lingua* L.
Q19	*Orchis militaris*
Q20	*Orchis purpurea* Hudson
Q21	*Dactylorhiza fuchsii* (Druce) Soó
Q22	*Platanthera chlorantha* (Custer) Reichb. f.
Q23	*Orchis morio*
Q24	*Orchis simia*
Q25	*Dactylorhiza romana*
Q26	*Serapias vomeracea* (Burm.) Briq.
Q27	*Dactylorhiza saccifera* (Brongn.) Soó
Q28	*Dactylorhiza fuchsii*
M10	*Ophrys vernixia*
Cc	—

The work with the rare British species highlighted the difficulties that can be encountered in developing effective symbiotic or asymbiotic methods, and revealed that though some species may respond very well to a number of fungal symbionts, or develop on a number of different media, no single method can be adopted for the *in vitro* propagation of all European orchid species, rather there are some which show a high degree of specificity for fungi and/or particular nutrients.

Screening tests for fungal compatibility

Before engaging in any large scale sowings to raise seedlings symbiotically it is necessary to establish that the orchid/fungus combination is going to be an effective and positive one. The easiest way to determine this is to run a brief, preliminary compatibility test.

Two screening tests were carried out, which revealed both their usefulness and the problems associated with a short term duration. Three species from each of the genera *Dactylorhiza*, *Ophrys*, *Orchis* and *Serapias* were screened for compatibility with nine vigorous orchid symbionts and eight less vigorous ones. Table 2 shows the source

Table 3. *Screening test of vigorous isolates: stages of development (see Figure 1 for explanation) after 2 weeks.*

Orchid species	Isolate								
	Q3	Q6	Q8	Q13	Q14	Q23	Q26	Q27	Cc
Dactylorhiza:									
D. elata (Poiret) Soó	0	0	0	0	0	0	0	0	0
D. fuchsii	0	1	1	0	0	1	0	0	1
D. praetermissa (Druce) Soó	0	0	0	0	0	0	0	0	0
Ophrys:									
O. apifera Hudson	0	0	1	0	0	0	1	0	0
O. holoserica	0	1	0	0	0	0	0	0	0
O. sphegodes	0	0	0	0	0	0	0	0	0
Orchis:									
O. laxiflora	2	2	2	2	2	3	2	2	3
O. militaris	0	0	0	0	0	0	0	0	0
O. morio	2	2	2	2	2	3	2	2	2
Serapias:									
S. lingua	2	2	—	0	0	2	2	0	2
S. neglecta De Not	0	0	—	0	0	2	0	0	0
S. vomeracea	2	1	—	0	0	3	2	0	2

species for the fungal isolates used. With the vigorous isolates, some orchid species responded very quickly, and so assessment was carried out after two weeks. With the less vigorous ones, response to infection was generally much slower, and seven weeks were allowed to elapse before assessment was carried out. After the assessments were made, the sowings were allowed to continue their development over a period of several months. This enabled the screening results to be viewed in relation to continued compatibility beyond the initial stages, through to development of mature seedlings which took place over this later period, and results are discussed accordingly.

Table 3 shows the results of the screening test over two weeks with the vigorous isolates. Isolate Q3, though not stimulating any of the *Dactylorhiza* or *Ophrys* species tested within two weeks, has in the long term been successful for raising *Ophrys* seedlings. The species of *Orchis* and *Serapias*, though reaching stage 2 quickly, developed no further.

With Q6 and Q8, species of *Orchis* and *Serapias* again developed to stage 2 within two weeks, but none of the species tested progressed beyond stage 3.

Q13, though initially stimulating the germination of *Orchis laxiflora* and *Orchis morio* up to stage 2 within two weeks subsequently only proved effective with *Orchis laxiflora*, and then only up to stage 4, over five months.

Q14 proved to be highly compatible with *Orchis laxiflora*, and subsequently with *Serapias lingua*, which was not stimulated within the test period. Mature seedlings have been obtained with both these species using Q14. However, *Orchis morio*, which reached stage 2 within two weeks, did not develop further. None of the *Dactylorhiza* species or *Ophrys* species germinated within the test period with Q14, but *Dactylorhiza fuchsii* did respond more slowly, and went on to reach stage 4 after two months.

Q23 was initially quick to stimulate the germination and development of *Orchis laxiflora*, *O. morio*, *Serapias lingua*, *S. neglecta* and *S. vomeracea*, and continued to be a compatible isolate for all these species throughout seedling growth. It is one of the isolates now used for routine micropropagation of a number of *Orchis* and *Serapias* species. Though initially stimulating *Dactylorhiza fuchsii* also, it had limited long term compatibility with the *Dactylorhiza* species tested, none of them progressing beyond stage 3.

Q26 initially stimulated species of *Ophrys*, *Orchis* and *Serapias*, though resulted in no further development of *Ophrys apifera*, *Orchis laxiflora*, *Serapias lingua* or *S. vomeracea*. *Orchis morio*, which reached stage 2 within the two weeks, went on to develop into mature seedlings over the following ten weeks.

Q27 stimulated *Orchis laxiflora* and *O. morio*, but no other species under test. However it proved to be ineffective, even for these two species, beyond stage 2.

With Cc, *Orchis laxiflora*, *O. morio*, *Serapias lingua* and *S. vomeracea* all showed a good response within the two weeks, but only *O. laxiflora* and *O. morio* went on to develop into mature seedlings. With *Orchis morio*, in particular, a good symbiosis was established, and protocorms reached stage 5 after about 7 weeks. This orchid/fungus combination has since been used in bulk sowings. The response with *Dactylorhiza* species did not go beyond stage 2.

These results show that in compatibility testing, even if there is a good initial response within only two weeks, this is not always indicative of subsequent compatibility throughout the later stages of seedling development. For example, Q3 initially appeared to be a compatible fungus for *Orchis* and *Serapias* species, but in fact turned out to be one of the most successful symbionts for *Ophrys* species, including *Ophrys holoserica*, the species from which it was isolated, and to be ineffective long term with *Orchis* and *Serapias* species.

Table 4 shows the results over seven weeks with the less vigorous isolates. Q17 stimulated species of all four genera tested, but protocorms did not eventually develop beyond stages 2 or 3.

Q18, isolated from *Serapias lingua*, showed the best germination response with *Serapias lingua*, and also subsequent compatibility up to mature seedling production.

Table 4. *Screening test of less vigorous isolates: stages of development (see Figure 1 for explanation) after 7 weeks.*

Orchid species	Isolate							
	Q17	Q18	Q19	Q20	Q21	Q22	Q24	Q25
Dactylorhiza:								
D. elata	1	—	—	0	0	0	2	2
D. fuchsii	0	0	0	0	0	0	0	0
D. praetermissa	0	0	0	1	0	2	1	0
Ophrys:								
O. apifera	1	0	1	1	0	0	1	0
O. holoserica	1	0	0	0	0	0	1	1
O. sphegodes	0	0	1	1	0	0	1	0
Orchis:								
O. laxiflora	2	2	2	2	4	2	4	1
O. militaris	0	0	0	0	0	0	0	0
O. morio	2	2	2	2	4	4	4	1
Serapias:								
S. lingua	1	3	2	2	1	0	4	1
S. neglecta	0	1	0	0	0	0	0	1
S. vomeracea	0	1	0	0	0	0	3	1

With the other species, protocorms did not progress beyond stages 1 and 2, as shown in the table.

Q19, though initially stimulating the germination of *Ophrys apifera*, *O. sphegodes*, *Orchis laxiflora*, *O. morio* and *Serapias lingua* , did not continue this compatibility into further development. However *Orchis militaris*, the source species for this isolate, did eventually form a successful symbiosis after three and a half months, and stage 5 protocorms were obtained after one year.

The same species responded with Q20 as with Q19. In addition, *Dactylorhiza praetermissa* was stimulated. But again, none developed beyond the initial stages, except for *Orchis militaris*,which went on the stage 4 within six months and stage 5 after 9 months.

Q21 appeared highly compatible with *Orchis laxiflora* and *Orchis morio*, though no positive response was obtained with the *Dactylorhiza* species within the seven weeks. Subsequently, all three species of *Dactylorhiza* germinated well with this isolate and went on to develop into mature seedlings over five months. The initial compatibility with *Orchis laxiflora* and *O. morio* was sustained, and seedlings developed fairly rapidly, within three months. This fungus is now used routinely for the propagation of these *Dactylorhiza* and *Orchis* species. The isolate was not successful with *Ophrys* or *Serapias* species.

Q22 seemed a promising symbiont at first for *Dactylorhiza praetermissa, Orchis laxiflora* and *O. morio*, but protocorms of all these species did not develop any further after the seven weeks, and fungal pathogenicity occurred.

Q24 was quick to stimulate the germination of *Orchis laxiflora, O. morio, Serapias lingua, S. vomeracea*, and to a lesser extent, the *Ophrys* species and two of the *Dactylorhiza* species tested. With *Ophrys* and *Dactylorhiza*, the germinated protocorms did not develop further than the stages reached in the test period, but with *Orchis* and *Serapias* species the compatibility lasted through protocorm development to the mature seedling stage. This isolate has since proved to be the most successful one so far for *Serapias* micropropagation.

With Q25, species from each of the genera tested germinated, but none progressed further than the stages shown in Table 4.

The major points to be noted from these screening tests are:

1. The orchid species will not always develop to the seedling stage with the fungi with which they initially germinate. For example, almost all the fungi tested showed some degree of compatibility with *Orchis laxiflora, O. morio* and one of the *Serapias* species, usually with *S. lingua*. However, compatibility beyond the initial stages only continued with a few isolates, and out of all 17 tested, only four proved to be useful for the routine propagation of *Orchis laxiflora*, three for *Orchis morio*, one for all the *Serapias* species tested and one for the *Dactylorhiza* species tested.

2. Sometimes an initial symbiosis may subsequently develop into a pathogenicity, with the fungus over-growing the protocorms, resulting in their death. This phenomenon has been observed and described fully by Hadley (1970).

3. The length of any test period must be adequate for the slower growing species, in order to determine whether or not a fungal isolate will establish a successful symbiosis, resulting in seedling production, rather than just early protocorm development. For example, over the test periods, *Ophrys* and *Dactylorhiza* species showed no response at all with the isolates which were eventually found to be the most suitable and effective symbionts.

4. The source species of the fungus, that is the orchid species from which it was isolated, cannot be used as a reliable guide to predict successful orchid/fungus combinations, though some degree of specificity is apparent, for example with *Ophrys* species and with *Orchis militaris*. The *Ophrys* species developed well with only one of the fungi tested, Q3, and this had been isolated from the roots of a plant of *Ophrys holoserica*. *Orchis militaris* developed eventually only with Q19 and Q20, isolated from *Orchis militaris* and the closely related *Orchis purpurea*, respectively.

Table 5. *Résumé of the furthest stages reached for 19 orchid species sown with 15 fungal isolates*

Orchid species	Isolate														
	Q3	Q6	Q8	Q14	Q17	Q18	Q19	Q20	Q21	Q23	Q24	Q26	Q28	M10	Cc
Dactylorhiza:															
D. elata	0	2	–	2	1	2	0	0	5	2	2	2	3	0	2
D. fuchsii	0	1	1	4	0	2	0	0	5	3	0	0	3	0	2
D. praetermissa	0	0	–	0	0	2	0	1	5	2	1	0	3	0	4
Ophrys:															
O. apifera	1	0	1	0	1	0	1	1	0	0	1	1	0	1	0
O. doerfleri Fleischm.	[3]	1	1	0	1	0	1	1	0	1	3	1	1	3	0
O. holoserica	4	1	0	0	1	0	0	1	0	0	1	0	0	1	0
O. insecifera L.	4	0	0	0	0	0	0	1	0	0	1	2	0	[3]	0
O. sphegodes	4	0	0	0	0	0	1	1	0	0	1	0	0	5	0
O. vernixia	1	0	0	0	0	0	0	1	0	0	1	0	1	4	0
Orchis:															
O. coriophora subsp. fragrans Poll. Sudre	2	2	3	5	3	5	2	2	5	[3]	5	[4]	3	2	4
O. laxiflora	2	2	3	5	3	2	2	2	5	5	5	2	3	3	5
O. militaris	0	0	2	0	2	0	5	5	0	0	0	0	2	0	2
O. morio	2	2	2	2	2	2	2	2	5	5	4	5	5	3	5
O. sancta	2	2	3	5	2	5	2	2	5	5	5	[4]	3	3	4
O. simia	2	2	2	2	3	2	3	4	2	3	3	2	2	2	2
Serapias:															
S. lingua	2	2	2	5	2	5	2	1	1	4	5	2	2	2	2
S. neglecta	1	1	0	1	2	1	1	[1]	0	3	[2]	0	2	2	1
S. parviflora Parl.	2	2	2	1	3	1	2	1	2	3	5	5	2	2	2
S. vomeracea	2	1	0	1	2	1	2	2	0	4	5	2	2	2	2

Values in parenthesis are for sowings under two months old, or incomplete testing due to contamination.

It has been possible to study this question of specificity more closely using a résumé of the most advanced stages so far reached, regardless of time taken, for 19 orchid species sown with 15 fungal isolates. Table 5 shows the development response over the whole range of orchid/fungus combinations. Most of the orchids tested, especially the *Orchis* and *Serapias* species, showed some positive response with several fungi.

With the exception of *Orchis militaris* , the *Orchis* species developed to stage 2 with all isolates. The response was less marked with *Dactylorhiza* and *Serapias* species. Although up to six isolates stimulated development to stage 3 or above, only Q21 was really effective for all three *Dactylorhiza* species tested, and only Q24 for all three *Serapias* species fully tested.

For *Ophrys* species, germination occurred with a wide range of fungi, and in no discernible pattern, but only a few combinations resulted in protocorms beyond stage 1. Q3 and M10 were particularly effective, though not to the same degree with each species. Mature seedlings of *Ophrys holoserica* and *Ophrys sphegodes* were obtained with Q3, isolated from *Ophrys holoserica*, and mature seedlings of *O. sphegodes* and *O. vernixia* were obtained with M10, isolated from *O. vernixia*. This indicates that there is some mycorrhizal specificity between *Ophrys* and other genera tested, and perhaps, though to a lesser extent, within the genus.

With *Orchis*, the commoner species *O. laxiflora*, *O. morio*, *O. coriophora* subsp. *fragrans* and *O. sancta*, were able to develop well with many more fungi than the rare *O. militaris* and *O. simia*. *O. coriophora* subsp. *fragrans* and *O. sancta* both developed to maturity with Q14, Q18, Q21, Q23 and Q24. Except for Q18, these fungi were also very effective symbionts with *O. laxiflora*. *O. laxiflora* also reached stage 5 with Cc, though somewhat less readily. *O. morio* shares two of the same symbionts, Q21 and Q23. In addition, Q28 and Cc both appeared more suitable with *O. morio* than with the other *Orchis* species tested. Stage 5 was reached with Q26, which was also a good symbiont with *O. sancta* and *O. coriophora* subsp. *fragrans*. The only isolates with which *Orchis militaris* reached stage 5 were Q19 and Q20, and these isolates were relatively ineffective with any of the other species tested apart from *O. simia*. *O. simia* appeared to be less selective than *O. militaris* for forming symbioses initially, but only with Q20 was the stage of leaf production reached.

The pattern of specificity in *Orchis* that has emerged is one of groups. *O. militaris* and *O. simia*, the rare species, are together in one group; *O. morio* is in a separate group, and this overlaps a third group containing the other three *Orchis* species. Into this third group two other separate groups, the *Dactylorhiza* group and the *Serapias* group, also overlap.

Summary. Taking into account the orchid/fungus combinations resulting in mature seedling production, with certain isolates there is a noticeable specificity between and within orchid genera, for example with the isolates Q3, Q20, Q28 and M10. With

Table 6. *Most suitable mycorrhizal fungus/orchid combinations for routine micropropagation to produce mature seedlings.*

Mycorrhizal fungus isolate	Orchid species source of isolate	Compatible orchid species
M10	*Ophrys vernixia*	*O. insectifera, O. sphegodes, O. vernixia.*
Q8	*Ophrys holoserica*	*O. holoserica, O. insectifera, O. sphegodes.*
Q14	*Orchis papilionacea*	*O. coriophora* subsp. *fragrans, O. laxiflora, O. sancta.*
Q20	*Orchis purpurea*	*O. militaris.*
Q21	*Dactylorhiza fuchsii*	*D. elata, D. fuchsii, D. praetermissa, D.romana, Orchis coriophora* subsp.*fragrans, O. laxiflora, O. morio, O. sancta.*
Q23	*Orchis morio*	*O. laxiflora, O. morio, O. sancta.*
Q24	*Orchis simia*	*O. coriophora* subsp.*fragrans, O. laxiflora, O.sancta, Serapias lingua, S. parviflora, S. vomeracea.*
Q28	*Dactylorhiza fuchsii*	*Orchis morio.*

others, Q14, Q18, Q21 and possibly Q23, there is a fair degree of non-specificity between orchid species. However, if one looks only at the early stages of development, up to stages 2 and 3, then a much higher degree of non-specificity is apparent. These results therefore illustrate the dangers of drawing premature conclusions from compatibility trials or sowing experiments conducted over short time periods. The results may also help to explain why some workers report non-specificity in symbiotic germination of European orchids (Hadley 1970) while others believe that the orchid/fungus relationship is fairly specific (Clements 1982).

Routine sowings and bulk propagation

Where consistently compatible orchid/fungus pairings have been found which continue to remain balanced throughout seedling development, it has been possible to use the symbiotic system routinely for large scale sowings, so that seedlings may be raised in bulk. Table 6 gives a summary of the most effective orchid/fungus combinations which are now used for the production of mature seedlings and for bulk propagation. The two most vigorous and effective isolates so far are Q21, derived from root tissue of *Dactylorhiza fuchsii*, which has proved an excellent symbiont with *Dactylorhiza* and *Orchis* species, and Q23, isolated from the root tissue of *Orchis morio*, and highly compatible with *Orchis laxiflora*, *O. morio* and *O. sancta*. These fungi are now being used to raise numerous batches of bulk sown *Orchis* and *Serapias* species, with some mature seedlings already in soil mix *in vitro* prior to transfer to pots (Figure 8). Re-introduction trials were piloted in 1986 using

Figure 8. Growth room, showing seedlings at all stages from sowings in Petri dishes to mature seedlings in soil mix *in vitro*.

Orchis laxiflora and *O. morio* seedlings and continued in 1987 with *O. laxiflora* plants raised from these bulk sowings. In 1988 the *O. laxiflora* seedlings which had been transplanted to a site at Wakehurst Place flowered successfully.

References

Clements, M.A. (1982). Developments in the symbiotic germination of Australian terrestrial orchids. In *Proc. 10th World Orchid Conference, Durban, 1981*, ed. J. Stewart & C.N. van der Merwe, pp. 269–73. Johannesburg: South African Orchid Council.

Clements, M.A., Muir, H. & Cribb, P.J. (1986). Preliminary report on the symbiotic germination of European orchids. *Kew Bull.*, **41**, 437–45.

Curtis, J.T. (1936). The germination of native orchid seeds. *Am. Orchid Soc. Bull.*, **5**, 42–7.

Hadley, G. (1970). Non-specificity of symbiotic infection in orchid mycorrhiza. *New Phytol.*, **69**, 1015–23.

Harvais, G. (1973). Growth requirements and development of *Cypripedium reginae* in axenic culture. *Can. J. Bot.*, **51**, 327–32.

Mead, J.W. & Bulard, C. (1975). Effects of vitamins and nitrogen sources on asymbiotic germination and development of *Orchis laxiflora* and *Ophrys sphegodes*. *New Phytol.*, **74**, 33–40.

Muir, H.J. (1987). Symbiotic micropropagation of *Orchis laxiflora*. *The Orchid Rev.*, **95**, 27–9.

Norstog, K. (1973). New synthetic medium for the culture of premature barley embryos. *In Vitro*, **8**, 307–8.

Host–fungus relationships in orchid mycorrhizal systems

Summary

The understanding of the host–fungus relationships in the mycorrhizas of orchids is important in relation to the application of symbiotic methods to seed germination and seedling development, and also for re-establishment in natural conditions either from seed or tissue culture as one of the contributions to conservation. Despite extensive progress in knowledge of the role of fungi in the early stages of germination, the inherent difficulty in the use of the symbiotic technique has inhibited its application.

Orchid fungi are extremely variable and relatively few root inhabitants are true mutualistic symbionts. The outcome of the relationship between the partners is a finely balanced one and many fungi isolated from orchid mycorrhizas may, after a period in culture, become incompatible with, or even pathogenic to, orchid protocorms.

Mutualistically symbiotic fungi enhance the nutrition of germinating seeds by the transference of carbon compounds. Photosynthetically active seedlings and mature plants, however, may be quite independent of their fungal partners. Evidence suggests that in conditions of nutrient stress the fungal partner may mediate in the movement of phosphate and/or nitrogen compounds, as in other mycorrhizal systems.

Introduction

Orchids, whether epiphytic or terrestrial, generally grow and thrive in conditions of nutrient impoverishment. As with other higher plants growing in similar environments this may only be possible by the association of fungi with the roots and other subterranean parts – *the mycorrhizal association.*

The understanding of the orchid–fungus relationship is important in relation to management and conservation, not only because of the potential application of symbiotic methods to the propagation of endangered orchid species but also in relation to the ecological characteristics associated with the natural occurrence of potentially symbiotic fungi and their colonisation of roots of mature orchids. The role of orchid fungi in the nutrition of seedlings and adult plants in nature could be critical in relation to ecological competition and the survival of re-introduced populations, particularly in conditions of stress. This chapter reviews the present state of our knowledge of the symbiotic relationships between orchid and fungus and describes a technique for the establishment of defined mycorrhizas in orchid tissue cultures. Such techniques offer opportunities for basic studies on symbiosis in mature plants, propagation by tissue culture and the establishment of fungus-protected rare and difficult native species. The need is stressed for further research to bridge the gap between pure knowledge and its application.

Patterns of infection

Early studies of the microscopic structure of orchid roots established the universal occurrence of endophytic fungi in a fairly consistent intracellular pattern. Variations in this pattern are reflected in the relatively sparse infection of many epiphytic (and some terrestrial) tropical orchids (Hadley & Williamson 1972), contrasting with the dense infection of most terrestrial north temperate species and the so-called 'saprophytic' or heterotrophic forms. The total dependence of the latter on their associated fungi can be related to the well-known phenomenon of heavy infection in symbiotically-germinated seed of orchids in general. The juvenile, heterotrophic phase and its dependence for success on the recognition of, and then infection by, specific fungi belies the seemingly simple structural relationship and the presumed mutualistic nature of the symbiosis.

Ultra-structural studies on protocorms and roots show that the interface between host-cell plasmalemma and fungal hypha is fairly uniform (Hadley 1975). Cytokinetic effects causing host cell nuclei to undergo endoreduplication of DNA regularly occur, while hyperplasia leading to slight morphological distortion of root tissues is also common, especially in tropical species (Hadley 1986).

The regularity of root infection in the Orchidaceae has led to a general assumption that the mycorrhiza fulfils a specific nutritional function in the adult plant. However, in view of the difficulty of obtaining mature, photosynthetically-active, non-infected tissue, there has been no comparable physiological study of orchid roots. In contrast, numerous studies have been made on symbiotically infected seed and microscopic protocorm stages. It is convenient to consider the root and seed germination phases separately.

Infection and physiological functions of roots

In the absence of any study on the physiology of orchid roots it is not possible to answer the question as to whether mycorrhizal infection enhances functions such as nutrient uptake, water transport and translocation. Similarly it is not known whether development of leaves, flowering, morphological changes or overwintering processes are in any way related to root infection. Studies on healthy and pathogenically-infected chrysanthemums, potatoes and tomatoes have shown that flowering and physiological stress have profound effects on infection and disease development (Busch & Edgington 1967; Pegg & Jonglaekha 1981; Pegg & Holderness 1984). The difficulty of obtaining non-infected material can be overcome in part by the use of selected systemic fungicides which inhibit the activity of the mycorrhizal endophyte and thus enable some comparisons in cultural conditions. An alternative strategy is the development of mature axenic plants and synthesised specific mycorrhizas.

North temperate orchids mostly undergo an annual cycle of renewal of roots and tubers, as well as their aerial parts. This cycle can be reproduced in culture using a cold treatment of 0–5 °C for 4–8 weeks following which leaf initials extend and new roots form, ultimately leading to the development of a replacement tuber. *Dactylorhiza majalis* subsp. *purpurella* (T. & T.A. Steph.) Soó & Morsby-Moore (*D. purpurella*), grown from seed in symbiotic culture, can be taken through several such cycles, developing into plants similar to those found in natural conditions (G. Hadley unpublished). Tissue culture-derived, symbiotically-infected plants also develop into photosynthetically active mature plants comparable to those originating from seed and the wild type (Beardmore & Pegg 1981).

Asymbiotically-grown protocorms provided with suitable nutrients will undergo development to form plantlets which, although usually smaller, are substantially similar to mycorrhizal ones. Photosynthetic activity, accumulation of starch reserves and subsequent development are apparently normal. Inoculation of mycorrhizal fungi leads to root infection without any significant difference in rate of development and, as a corollary, the application of the systemic fungicide thiabendazole will arrest the fungus but has no phytotoxic effects on non-infected plantlets.

A limited study of two-year-old infected plantlets of *D. majalis* (G. Hadley unpublished) made comparisons of weight, root length, shoot length and infection pattern, using a drench treatment with thiabendazole in pot cultures to inhibit the fungus. There was no significant difference between untreated and treated plantlets. Both developed new roots which in the fungicide-treated plants were uninfected but were otherwise similar to infected ones. In these experiments the plants were not stressed.

Related work with *Goodyera repens* Br. (Alexander & Hadley 1985) showed that the relative growth rates of populations of both small and large (approx. 140 and 440 mg fresh weight respectively) mycorrhizal plants were unaffected by the presence of cellulose as an external carbon source, nor was there evidence of any movement of carbon (i.e. neither ^{14}C–carbohydrate from substrate via fungus to host, nor ^{14}C-CO$_2$, via host to fungus) between the partners. Experiments in which thiabendazole was applied to plants to reduce the mycorrhizal effect showed, again, that there were no apparent differences between mycorrhizal and non-mycorrhizal plants at this stage.

The fact that young plantlets of *G. repens* up to approx. 50 mg fresh weight, bearing one or two small leaves and root initials, take up carbohydrate from the substrate like protocorms clearly indicates that, at some stage, there is a transition from dependence on the fungus to independence associated with autotrophy.

Other studies by the same group of workers have investigated uptake of phosphate by culture-grown plants of *G. repens* (Alexander & Hadley 1984; Alexander *et al.* 1984). This work has clearly demonstrated, in conditions of nutrient stress, that the movement of phosphate into mycorrhizal roots was greater than that into non-mycorrhizal ones. Translocation of ^{32}P into roots through external mycelium was demonstrated and could be inhibited using thiabendazole. Moreover, relative growth rates of plants were enhanced by mycorrhizal activity and were reduced when such activity was inhibited by fungicide treatment. Uptake of nitrogen was also enhanced.

Symbiotic germination

The classical work on interpretation of the significance of orchid fungi in seed germination is well known and has been frequently reviewed (see Hadley 1982; Harley & Smith 1983). Since the early studies of Bernard (1909) it has been a general hypothesis that, following infection by a compatible fungus; the orchid seed is stimulated to germinate. Nevertheless there is always some increase in size of the embryo due to water absorption, before infection is possible, i.e. the first stages of germination must precede infection. This is clearly shown by observations of cleared material, e.g. in *G. repens* (Alexander & Hadley 1983). Similar observations have been made using other material (G. Hadley unpublished).

It should be noted, however, that the term 'germination' is used by some workers to describe sequential stages in the process of growth of protocorms. As a means of quantifying the assessment of such studies, germination is conveniently recognized as several stages (0–5), culminating in the development of leaf initials and photosynthetic tissue.

Clements & Ellyard (1979) regenerated interest in the symbiotic germination method using techniques developed by Warcup (1973; and see also Warcup 1981b) on Australian orchids. Success of the new technique, which is essentially the refinement of methods of isolation and selection of culture media, has led to its application to many north temperate species (Clements *et al.* 1986).

Significance of fungi in the protocorm stage

Regardless of whether infection causes germination or succeeds it, the stimulation of development following infection is known to be associated with the provision by the fungus of specific or more general nutrients, particularly carbon compounds, until the heterotrophic stage is over. The corollary, the assumption that in natural conditions such fungi merely provide nutrients, led to much work on asymbiotic culture and its applications (discussed elsewhere in this Volume) particularly for the commercial culture of orchids.

As well as providing nutrients, it seems likely that fungi transfer other, more specific metabolites such as enzyme precursors. For instance, asymbiotically grown protocorms readily take up nutrients and slowly accumulate starch, but appear unable to metabolise it for growth. Infection, in north temperate orchid protocorms at least, appears to initiate metabolic activity leading to the utilisation of accumulated starch reserves which would otherwise remain unused (Purves & Hadley 1976). Chlorophyll content and photosynthetic activity are enhanced in some tropical orchids (Hadley 1982). This suggests that processes associated with gene amplification or some other, as yet unknown, mechanisms of triggering latent metabolic activity may be involved.

Nevertheless it is still unclear as to whether the symbiotic growth stimulus (see later) results from a biotrophic movement (i.e. across a living interface) of metabolites from fungus to host or whether the process is a purely necrotrophic one involving the breakdown of the intracellular fungal hyphae by the metabolic activities of the host, i.e. the 'digestion' of the fungus.

The sequence of changes in hyphae and mycelial form following initial infection until the ultimate disappearance of pelotons is indicative of enzymic lysis. Hyphal lysis is a common and well-documented phenomenon in saprophytic fungi (Matile 1969) and in higher plants infected with pathogenic fungi (Burges 1939; Pegg & Vessey 1972; Pegg 1976). Autolysis, depending on the specific composition of the hyphal wall, involves the activity of a range of hydrolytic enzymes such as glucanases, chitinases and proteases.

Work on mycorrhizal and non-mycorrhizal tissue-cultured *Dactylorhiza majalis* showed that the orchid most probably plays an active role in the continuous restriction and elimination of the symbiont. Non-mycorrhizal protocorms were shown to contain the constitutive enzymes, $1,3-\beta-$glucanase endochitinase and N-acetylglucosaminidase capable of hydrolysing the fungal cell walls of *Thanatephorus*, *Ceratobasidium* and *Tulasnella*. When protocorms were infected with these symbionts the levels of the enzymes rose substantially (G.F. Pegg & B.J. Beardmore unpublished).

Orchid fungi and specificity

The fungi which occur in symbiosis with orchid roots are mainly non-sporing members of the imperfect genus *Rhizoctonia* or imperfect stages of Basidiomycetes. They can be readily isolated and may be grown in culture with comparative ease. In general, they do not require special nutrients and a simple growth medium based on sucrose and yeast (Clements 1982) is adequate. Growth on such a medium probably reflects their requirement for small amounts of vitamins or amino acids (Hadley & Ong 1978). Outside the orchid, *Rhizoctonia* spp. probably exist as non-sporing saprophytic mycelium, but little is known about the biology of these fungi in soil.

The first isolations were made by Noel Bernard (1909) who named three species, i.e. *Rhizoctonia repens, R. lanuginosa* and *R. mucoroides*. Since then, numerous rhizoctonias have been isolated, cultured, described and named as orchid fungi. Not all have been shown to be symbiotic however and few have survived in culture. Rhizoctonias from soil, and some originating as pathogens of crop plants, can also enter into symbiotic relationships with at least some north temperate orchids (Hadley 1970a). In suitable conditions of culture, many fungi from non-orchid hosts or from widely different geographical sources stimulated a growth response in seed germination tests. The range was very great, from slow growing, benign or incompatible isolates, through a broad spectrum of differing degrees of symbiotic activity to aggressive isolates which readily developed a pathogenic phase in appropriate conditions. In the latter case, usually no more than a very small percentage of the seed population survived although the survivors were vigorous and healthy.

The situation is complicated by the physiological and genetic variability of such fungi, and the fact that they may not retain symbiotic activity after a period in artificial culture. For example, Alexander & Hadley (1983) found that among 30 isolates from *Goodyera repens*, all were recognisable as the same fungus in appearance and cultural characteristics. Nevertheless there was a wide range of symbiotic activity, as measured by the growth response of protocorms. At the same time many of the infected protocorms in a population failed to develop any faster than non-infected controls, resulting in a bimodal population containing a small proportion of very large protocorms.

Isolates which had been in culture for one or two years, or were re-isolated from cultured protocorms (or plantlets) after the same period, lost some of their ability to stimulate growth and showed more variability than when originally isolated. It is likely that such fungi are unstable in culture, as is known to be the case with other rhizoctonias such as strains of the pathogenic *Rhizoctonia solani*.

Because of their tremendous variability and the difficulty of recognising individual isolates it is not surprising that the orderly classification of orchid fungi is difficult if not virtually impossible. Many isolates of *Rhizoctonia* are now known to be anamorphic species of the Basidiomycete genera *Tulasnella, Sebacina* and *Ceratobasidium* (Warcup 1981a) (Table 1). Some species of *Thanatephorus* (including strains of aggressive pathogens better known as *Rhizoctonia solani*) have also been recognised as orchid fungi, or can be so in culture. But it is now clear that many such fungi occur on and inside orchid roots as casual associates which are not necessarily beneficial to their hosts.

Clearly, only fungi which are isolated from intracellular coils should be regarded as true endophytes. Among these, Warcup (1981b) recognises a degree of specificity amongst host-groupings. Clements *et al* . (1986) obtained 80 fungal isolates from 32 orchid taxa of which over 20 taxa responded to infection in culture tests. Specificity

Table 1. *Teleomorphic and anamorphic forms of Basidiomycete endophytes infecting* Dactylorhiza majalis.

Endophyte	Strain	Source
Thanetephorus cucumeris	W48	Soil, pathogen
	0269	Orchid endophyte
	W82	Soil, non-pathogen
	Rs1	Tomato pathogen
	T35	Orchid endophyte
	W87	Soil, non-pathogen
T. sterigmatus	060	Orchid endophyte
	0760	" "
Tulasnella calospora	Pb 47	" "
	0388	" "
	AMo4	" "
	062	" "
	0582	" "
	0689	" "
Ceratobasidium cornigerum	0479	" "
	0393	" "
	AD14	" "
	0167	" "
Ceratobasidium spp.	F1	" "
	T	" "
Rhizoctonia solani	Rs10	Rice pathogen

was evident in that *Ceratobasidium*-like strains related to *Dactylorhiza* spp. and *Tulasnella*-like strains related to *Orchis*, *Ophrys* and *Serapias* spp. But the concept of specificity is clearly related more to ecological characteristics than the original one host/one fungus theory of Burgeff (1909), and the wide taxonomic range of compatible symbiotic genera on one orchid species is evident (Table 1).

Finally the presence of Basidiomycete fungi such as *Armillaria* and *Fomes* in the roots of achlorophyllous ("saprophytic") orchids such as *Gastrodia* and *Galeola* spp. should be mentioned. Knowledge of these associations is very inadequate although a limited amount of work on seed germination and development has been carried out by Warcup (1981a) and Tashima *et al.* (1978).

Physiological aspects of the symbiotic growth stimulus

The rates of growth and development of asymbiotic material, even in culture conditions where uninfected seeds and non-infected protocorms were adjacent to symbiotically-infected ones, have invariably been much less than those of symbiotic material, regardless of the nutrient status of the medium. The consistency of the pattern of infection and development, independently of the species of fungus or host, and the fact that potentially pathogenic fungi can behave as suitable symbionts (Hadley 1982) argues for a specific functional role. Microscopy of populations of protocorms of *Dactylorhiza majalis* (Hadley & Williamson 1971) showed that an increase in length or volume was recognizable within one or two days of initial penetration by the fungus and within a few hours of the first appearance of intracellular coils (Figure 1). Growth continued to accelerate and development of protocorms was dramatically enhanced by infection, in north temperate orchids as in several tropical species which also responded within a few days of infection (Hadley 1982). For instance *Spathoglottis plicata* (Figure 2), which contains chlorophyll from the earliest stages of germination, is extremely responsive to infection whereas non-infected protocorms grow slowly regardless of nutritional and environmental factors.

Following some post-germination development, both asymbiotic and mycorrhizal seedlings progress to the plantlet stage. Combinations of sugars, growth factors and undefined organic materials are necessary to encourage asymbiotic cultures, and normal commercial practice follows this pattern. Even the achlorophyllous orchid *Galeola septentrionalis* Reichb. f. has been grown using suitable cytokinins

Figure 1. Stimulation of growth (increase in length) of protocorms of *Dactylorhiza majalis* subsp. *purpurella* in response to infection by a mycorrhizal endophyte, *Rhizoctonia* isolate Rs 10.

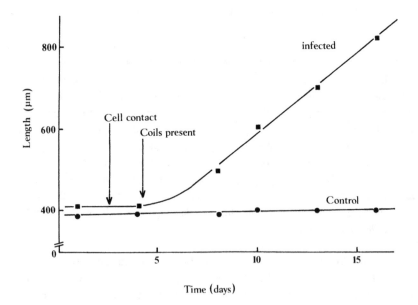

(Nakamura 1982) in this way to produce plantlets with 10 mm shoots, similar to the heterotrophic plantlets of north temperate species (Hadley 1970b).

Mycorrhizal seedlings, as has been indicated, grow more rapidly than asymbiotic ones, and, since their fungi have simple nutrient requirements they need only a cellulose carbon source. Translocation of metabolic material into the orchid, and transfer from fungus to host by biotrophic or necrotrophic processes, is the primary source of nutrients, especially carbon compounds, during the heterotrophic and juvenile autotrophic phases of development.

In *Goodyera repens*, as previously mentioned, carbon transfer from fungus to orchid switches off as photosynthetic activity becomes enhanced following leaf production (Alexander & Hadley 1985). Provided that mineral nutrients are adequate, as in horticultural practice, seedlings may thus become largely independent of symbiotic fungi as they develop. The same is probably true of adult plants. The physiological stimulus to growth may therefore be a feature only of the germination phase. Nevertheless, much needs to be done in observing, isolating and characterizing orchid fungi to determine the extent to which they represent a resource for use in symbiotic germination.

Figure 2. Response of *Spathoglottis plicata* to infection by a strain of *Tulasnella calospora*, isolate AMo4, in darkness (■) compared with growth of non-infected protocorms on glucose-salts nutrient medium in conditions of light (●) and darkness (▲).

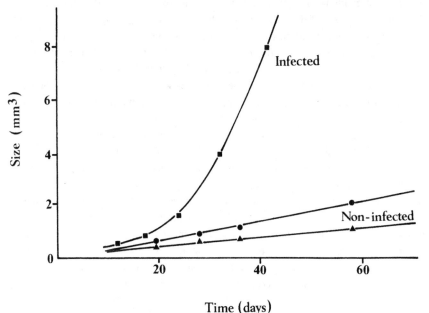

Nutritional patterns in mature orchids

Since there appears to be no movement of photosynthetically fixed carbon from host to fungus in adult orchids, their mycorrhizas may therefore be unique in that carbon moves from fungus to host only in seedling stages, except in the totally achlorophyllous holosaprophytes. On the other hand, the enhanced uptake of phosphate in stress conditions is comparable with the situation in other mycorrhizas and may be normal for orchids in field conditions. This new concept for orchid mycorrhizas needs to be evaluated in relation to tropical orchids where the available phosphorus in the substrate may be a critical factor for growth.

It seems possible therefore that the function of orchid mycorrhizal fungi in adult plants may compare with that in other mycorrhizas such as vesicular-arbuscular and ectotrophic (sheathing) types, to mediate in uptake of mineral nutrients in stress conditions.

Control and recognition

One of the most important, yet little understood, aspects of the orchid-fungus symbiosis is the mechanism(s) of integration which enable the recognition of compatible fungi at the root surface, the inhibition of incompatible or potentially pathogenic strains of similar species and, in the case of tubers and other fleshy storage organs, the control of a variety of putative pathogens.

Orchid fungi always penetrate root hairs or other epidermal cells by single hyphae, by way of infection pegs, and then extend inside the cell to form a close interface with the living cytoplasm of their host. Similar fungi which are pathogens of other hosts and normally infect by forming dense infection cushions will, in culture with orchid roots, infect them by way of single hyphae and, clearly, are subject to the controlling mechanism operated by the host. Subsequently, the intracellular development of such fungi follows a uniform pattern with clear features which vary only if and when the relationship breaks down due to the fungus either (i) being inhibited or (ii) becoming aggressive leading to 'breakaway parasitism' (Hadley 1982).

As with plant–pathogen interactions nothing is known about the recognition mechanism, although resistance to phytoalexin-like antifungal substances has been implicated in the establishment and localisation of infection. The role of phytoalexins in resistance mechanisms to some pathogenic fungi has been well documented (Bailey & Mansfield 1982). In orchids the phenanthrene antifungal compounds, orchinol, loroglossol and hircinol show many of the properties of phytoalexins. They were characterised as intermediate metabolites of tubers and bulbs and presumably roots. Their precise role in the exclusion of non-symbionts of orchids or the tolerance of them by symbionts in orchid mycorrhizas in general is not clear. A strain of *Ceratobasidium* sp. established as a symbiont in *Dactylorhiza majalis* subsp. *purpurella* was able to grow unaffected in concentrations of up to 150μg orchinol, well in excess of levels reported *in vivo*. A pathogenic strain of *Thanatephorus cucumeris* responded similarly. Orchinol appeared to be unstable in culture, falling to 60% of its original concentration after 7 days in the presence or absence of the fungus (G.F. Pegg and B.J. Beardmore unpublished).

The establishment of mycorrhizal infection in axenically-grown orchids

Based on their early studies on *Cymbidium*, Bernard (1909) and Burgeff (1909) considered that orchid seed was incapable of germination in the absence of a symbiont. This concept was disproved by Bernard (1911) himself and by the pioneering work of Knudson (1922) who demonstrated that an extracellular source of carbohydrate, fructose, would substitute for the symbiont. Although Knudson's methods were widely adopted by orchid growers the development of tissue culture *per se* did not occur until 40 years later with the work of Wimber (1963) and Morel (1964).

Preliminary experiments explored different aqueous culture media to maximise the quantity of axenic orchid tissue. When microscopic seed-derived protocorms of *Dactylorhiza majalis* are grown as non-differentiating cultures, large, swollen, rhizoidal-like masses develop. These are not protocorms *sensu stricto* and have been termed protocorm-like. Growth of this tissue was slow on solid media or in deep submerged aqueous culture. Routine production of axenic tissue was obtained by culturing aseptically 6, 5–10 mm protocorm segments in 10 cm^3 of an orchid culture medium in a wool-plugged 100 cm^3 conical flask in darkness at 22 °C (Beardmore & Pegg 1981).

Early attempts to obtain mycorrhizal infection by inoculating strains of the symbionts listed in Table 1 on to aqueous orchid tissue cultures or protocorms transferred to an agar medium were unsuccessful. With few exceptions the fungal strains preferentially utilised the 2% sucrose and other organic constituents of the culture medium as the most readily available source of carbon. Mycelium overgrew the orchid tissue resulting in a killing of the tissue. Similarly, if fungal strains encountered a cut protocorm surface no symbiosis occurred and the reaction was pathogenic, with browning and tissue disintegration.

Mycorrhizal synthesis was achieved by growing orchid explant and symbiont in juxtaposition but on different media (Figure 3). Slopes of 5 cm^3 of 2% cellulose-Pfeffer agar were prepared in 100 cm^3 urethane-stoppered conical flasks. A second medium, 10 cm^3 0.4% water agar was added aseptically as a bottom layer. An entire protocorm explant was transferred to the tap water agar and a strain of *Rhizoctonia* from a *Ceratobasidium* sp. was inoculated in the centre of the cellulose-Pfeffer slope. In the absence of nutrients in tap water agar to sustain orchid growth, mycorrhizal establishment was indicated by a dry weight increase in the protocorm based on carbon from hydrolysed cellulose translocated by the symbiont (Table 2). Infection occurred between 14 and 21 days after culture establishment. Orchid tissue changed

Table 2. *The effect of* Rhizoctonia *mycorrhizal infection on dry weight increase in tissue cultured protocorm of* Dactylorhiza majalis.

	Dry wt orchid protocorm (mg) (means of 10 replicates)		
Age of culture (d)	Non-inoculated	*Rhizoctonia* strain T-inoculated	Difference from non-inoculated	LSD of means
0	18.0	20.3	+2.3	5
7	17.6	21.3	+3.7	6.1
14	16.9	15.4	−1.5	6.8
21	19.1	26.9	+7.8	6.8*
28	16.2	31.8	+15.6	13.6***

*, ***, Results significantly different.
P = 0.05 and 0.001 respectively.

Figure 3. Double culture technique for the establishment of orchid mycorrhizal tissue using axenically-grown protocorms.

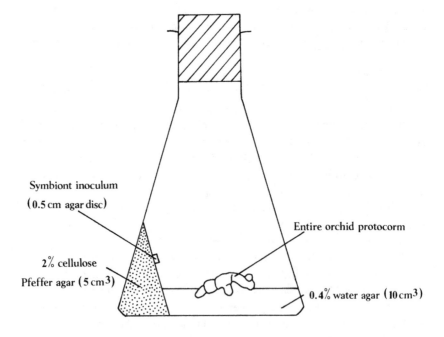

Symbiont inoculum
(0.5 cm agar disc)

Entire orchid protocorm

2% cellulose
Pfeffer agar (5 cm³)

0.4% water agar (10 cm³)

from a translucent appearance to opaque and showed a 96% increase in dry weight after 3 weeks. Sections through the tissue showed dense peloton formation in all stages of establishment and lytic disintegration. The cellulose medium did not support excessive hyphal growth and its hydrolysis presumably depends on the induction of cellulase and cellobiase.

The extreme variation found in the reaction of *Dactylorhiza* to endophyte infection in relation to strain source, temperature, medium composition and inoculation procedure, emphasises the fine balance that exists between the situation where the fungus is pathogenic on the orchid and that where the orchid is able to contain and destroy the fungus. These experiments emphasise that nutrition, especially the source and availability of carbon, is probably the determining factor in the equation. *Rhizoctonia* spp. are among the most destructive of plant pathogens and it may not be incidental to the mycorrhizal story that the main agents of pathogenesis in *Rhizoctonia* diseases, the pectolytic and cellulolytic enzymes, are closely controlled in induction and repression by the type and concentration of available sugars.

The importance of mycorrhizas in the management and conservation of orchids may extend beyond nutritional considerations. The phenomenon of cross-protection in higher plants is now well established, where infection by an avirulent pathogen precludes infection and subsequent disease development by a virulent one. This has been described in ectotrophic mycorrhizas by Marx (1973) and almost certainly exists in orchid and other endotrophic associations. Axenically-grown orchids would be especially vulnerable to disease if transferred as non-mycorrhizal plants to the wild or to another non-sterile growing medium.

Conclusions

The management and conservation of orchids requires a complex strategy involving ecological and environmental balances, the preservation of natural habitats and importantly an understanding of the whole biology of the endangered species. The contribution of research on mycorrhizal associations is fundamental to the success of this strategy, particularly where species derived from seed or tissue culture require to be reintroduced into native habitats, or for a full understanding of the nutritional requirements of wild species.

Most research to date has been conducted on a limited number of genera of temperate terrestrial orchids. Research is urgently required on some of the endangered species, the very nature of the problem limiting the availability of experimental material. Similarly there are profound gaps in our knowledge of tropical species both epiphytic and terrestrial. Such research is complementary to other aspects of orchid conservation. The important requirement for the future is the careful integration of scientific research into the applied development phase of the strategy, since seed and tissue culture methods of propagation cannot be bypassed.

References

Alexander, C., Alexander, I.J. & Hadley, G. (1984). Phosphate uptake in relation to mycorrhizal infection. *New Phytol.*, **97**, 401–11.

Alexander, C. & Hadley, G. (1983). Variation in symbiotic activity of *Rhizoctonia* isolates from *Goodyera repens* mycorrhizas. *Trans. Br. Mycol. Soc.*, **80**, 99–106.

Alexander, C. & Hadley, G. (1984). The effect of mycorrhizal infection of *Goodyera repens* and its control by fungicide. *New Phytol.*, **97**, 391–400.

Alexander, C. & Hadley, G. (1985). Carbon movement between host and mycorrhizal endophyte during the development of the orchid *Goodyera repens* Br. *New Phytol.*, **101**, 657–65.

Bailey, J.A. & Mansfield, J.W. (ed.) (1982). *Phytoalexins*. New York: Wiley.

Beardmore, Jane & Pegg, G.F. (1981). A technique for the establishment of mycorrhizal infection in orchid tissue grown in aseptic culture. *New Phytol.*, **87**, 527–35.

Bernard, N. (1909). L'évolution dans la symbiose, les orchidées et leurs champignons commensaux. *Ann. Sci. Nat. Bot. Sér. 9*, **9**, 1–196.

Bernard, N. (1911). Sur la fonction fongicide des bulbes d'Ophrydees. *Ann. Sci. Nat. Bot. Sér. 9*, **14**, 221–34.

Burgeff, H. (1909). *Die Wurzelpilze der Orchideen, ihre Kultur und irh Leben in der Planze*. Jena: G. Fisher.

Burges, A. (1939). The defensive mechanism in orchid mycorrhiza. *New Phytol.*, **38**, 273–83.

Busch, L.V. & Edgington, L.V. (1967). Correlation of photoperiod with tuberisation and susceptibility. *Can. J. Bot.*, **45**, 691–3.

Clements, M.A. (1982). Developments in the symbiotic germination of Australian terrestrial orchids. In *Proc. 10th World Orchid Conference, Durban, 1981*, ed. J. Stewart & C.N. van der Merwe, pp. 269–73. Johannesburg: South African Orchid Council.

Clements, M.A. & Ellyard, R.K. (1979). The symbiotic germination of Australian terrestrial orchids. *Am. Orchid Soc. Bull.*, **48**, 810–6.

Clements, M.A., Muir, H.J. & Cribb, P.J. (1986). A preliminary report on the symbiotic germination of European terrestrial orchid species. *Kew Bull.*, **41**, 437–45.

Hadley, G. (1970a). Non-specificity of symbiotic germination in orchid mycorrhiza. *New Phytol.* **69**, 1015–23.

Hadley, G. (1970b). The interaction of kinetin, auxin and other factors in the development of north temperate orchids. *New Phytol.*, **69**, 549–55.

Hadley, G. (1975). Organization and fine structure of orchid mycorrhiza. In: *Endomycorrhizas*, ed. F.E. Sanders, B. Mosse & P.B. Tinker, pp. 335–51. London: Academic Press.

Hadley, G. (1982). Orchid mycorrhiza. In *Orchid Biology: Reviews and Perspectives*, Vol. II, ed. J. Arditti, pp. 83–118. Ithaca, N.Y.: Cornell University Press.

Hadley, G. (1986). Mycorrhiza in tropical orchids. In *Proc. Fifth ASIAN Orchid Congress Seminar, Singapore, August 1–3, 1984*, ed. A.N. Rao, pp. 154–9. Singapore: Parks & Recreation Dept., Ministry of National Development.

Hadley, G. & Ong, S.H. (1978). Nutritional requirements of orchid endophytes. *New Phytol.*, **81**, 561–9.

Hadley, G. & Williamson, B. (1971). Analysis of the post-infection growth stimulus in orchid mycorrhiza. *New Phytol.*, **70**, 445–55.

Hadley, G. & Williamson, B. (1972). Features of mycorrhizal infection in some Malayan orchids. *New Phytol.*, **71**, 1111–8.

Harley, J.L. & Smith, S.E. (1983). *Mycorrhizal Symbiosis*. London: Academic Press.

Knudson, L. (1922). Non symbiotic germination of orchid seeds. *Bot. Gaz.*, **73**, 1–25.

Marx, D.H. (1973). Mycorrhizae and feeder root diseases. In: *Ectomycorrhizae, Their Ecology and Physiology*, ed. G.C. Marks & T.T. Kozlowski, pp. 351–77. New York & London: Academic Press.

Matile, P.H. (1969). Plant lysosomes. In *Lysosomes in Biology and Pathology 1*, ed. J.T. Dingle & H.B. Fell, pp. 406–30. Amsterdam: North Holland Publishing Co.

Morel, G.M. (1964). Tissue culture – a new means of clonal propagation of orchids. *Am. Orchid Soc. Bull.*, **33**, 473–8.

Nakamura, S.J. (1982). Nutritional conditions required for the non-symbiotic culture of an achlorophyllous orchid *Galeola septentrionalis*. *New Phytol.*, **90**, 707–15.

Pegg, G.F. (1976). Glucanohydrolases of higher plants: a possible defence mechanism against parasitic fungi. In *Cell Wall Biochemistry Related to Specificity in Host–Plant Pathogen Interactions*, ed. B. Solheim & J. Raa, pp. 305–47. Universitetsforlaget, Tromso–Oslo–Bergen.

Pegg, G.F. & Holderness, M. (1984). Infection and disease development in NFT-grown tomatoes. In *Proc. VIth Int. Congr. on Soilless Culture, Lunteren*. pp. 493–510. Wageningen: Secretariat ISOSC., P.O. Box 52, 6700 AB Wageningen, The Netherlands.

Pegg, G.F. & Jonglaekha, N. (1981). Assessment of colonisation in *Chrysanthemum* grown under different photoperiods and infected with *Verticillium dahliae*. *Trans. Br. Mycol. Soc.*, **76**, 353–60.

Pegg, G.F. & Vessey, J.C. (1972). Chitinase activity in *Lycopersicon esculentum* and its relationship to the *in vivo* lysis of *Verticillium alboatrum* mycelium. *Physiol. Plant Pathol.* 3, 207–22.

Purves, S. & Hadley, G. (1976). The physiology of symbiosis in *Goodyera repens*. *New Phytol.*, 77, 689–96.

Tashima, Y., Terashita, T., Umata, H. & Matsumoto, M. (1978). *In vitro* development from seed to flower in *Gastrodia verrucosa* under fungal symbiosis. *Trans. Mycol. Soc. Japan*, **19**, 449–53.

Warcup, J.H. (1973). Symbiotic germination of some Australian terrestrial orchids. *New Phytol.*, 72, 387–92.

Warcup, J.H. (1981a). Orchid mycorrhizal fungi. In: *Proc. Orchid Symposium, Sydney, 1981*, pp. 57–63. New South Wales Orchid Society.

Warcup, J.H. (1981b). The mycorrhizal relationships of Australian orchids. *New Phytol.*, **87**, 371–81.

Wimber, D. (1963). Clonal multiplication of Cymbidiums through tissue culture of the shoot meristems. *Am. Orchid Soc. Bull.*, **32**, 105–7.

The effects of the composition of the atmosphere on the growth of seedlings of *Cattleya aurantiaca*

Introduction

Considerable interest attaches to the control of growth and development of orchid protocorms and seedlings. A series of experiments were conducted which compared the relative merits of different culture vessels for the germination and growth of seedlings of *Cattleya aurantiaca* and then attempted to identify some of the changes which occurred in culture vessels during growth.

Materials and methods

Procedure used to surface-sterilize and sow seed

Seed was surface-sterilized using 5% commercial bleach (Domestos: Lever Bros., UK) for 1.5 minutes before sowing onto Thompson's medium (Thompson 1977). All vessels were placed in a Warren Shearer growth cabinet at a fluence rate of 142 μmol m^{-2} s^{-1}, a temperature of 22.5 ± 2 °C and a relative humidity of 90% (Seaton & Hailes 1989). Continuous light was used unless otherwise stated.

Measurement of growth and development

The percentage of seeds germinated was recorded at 14 days and 28 days, and growth was monitored at intervals by measuring the diameter of 50 protocorms, in each of 4 flasks, using a Leitz inverted microscope equipped with an eyepiece graticule. As this parameter gave no information about the development of protocorms an index of development was also employed, which was modified from that of Spoerl (1948). Seedlings were assigned to one of four different developmental stages (Figure 1). The number in each class was multiplied by the class number, and the values for the different classes summed to give the Protocorm Development Index (PDI). In some experiments the fresh and dry weights of 100 seedlings randomly selected from each flask was determined at the end of the experiment.

Vessels used to germinate and grow seedlings

Germination and protocorm growth were compared in seven different types of culture vessel as indicated below:

1. 25cm^3 'Sterilin' tissue-culture flasks.
2. 'Sterilin' 75 mm × 12 mm test-tubes.
3. 2.54 cm wide boiling tubes plugged with cotton bungs covered with polythene.
4. Petri-dishes sealed for seven-eights of their circumference with cellotape.
5. 125 cm^3 Erlenmeyer flasks with rubber bungs.
6. 125 cm^3 Erlenmeyer flasks with rubber bungs fitted with glass breather tubes plugged with cotton wool.
7. 125 cm^3 Erlenmeyer flask plugged with a cotton bung covered with polythene.

Figure 1. The four designated stages of protocorm development.

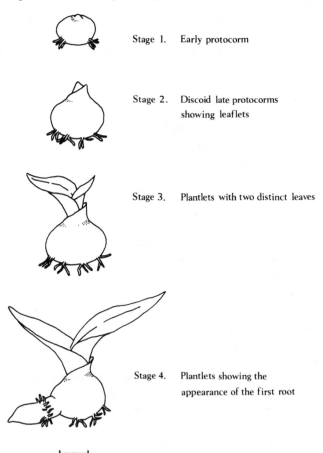

Stage 1.   Early protocorm

Stage 2.   Discoid late protocorms showing leaflets

Stage 3.   Plantlets with two distinct leaves

Stage 4.   Plantlets showing the appearance of the first root

500 µm

Analysis of the flask atmosphere

A Pye 104 (model 134) gas chromatograph with a katharometer detector was fitted with a 15.24×0.64 cm column packed with a molecular sieve (Linde 5A, 50–70 mesh) for the estimation of oxygen. A 2.54×0.64 cm column packed with silica gel was used for the measurement of carbon dioxide. However for the estimation of the concentration of ethylene the gas chromatograph was equipped with a flame ionisation detector and an alumina/sodium hydroxide (70–100 mesh) column.

Apparatus used to flush flasks continuously with specific concentrations of carbon dioxide and oxygen

Gas cylinders containing either 0.03, 1.0, 10.0 or 20.0% carbon dioxide and 21 or 10% oxygen were prepared by the British Oxygen Company. All of the cylinders were fitted with regulators preset at 2.07×10^5 Pa and flow meters which had been modified by Wescol Ltd. Four replicate flasks were employed for each treatment and each of the four flasks were connected to a central sterile flask containing sterile deionised water, through which incoming gases were bubbled to increase the water vapour content of the gases entering the culture flasks. The flow rates through the individual culture flasks were equalised to four air changes per hour using aquarium taps which were fitted to each flask. All flasks and tubing were autoclaved and assembled in a laminar flow hood before connection to the gas cylinders. Swinney filters were incorporated into the gas lines immediately after the flow metes and these filters were changed at 14 day intervals throughout the experimental period. Figure 2 shows a diagram of the apparatus used.

Results

In an experiment designed to examine germination and development in different culture vessels there was little or no difference in the percentage germination (i.e. all around 60%). However, within 70 days from the start of germination seedlings in flasks fitted with solid bungs, screw-capped tissue culture flasks and flat sided 'Sterilin' tubes showed markedly better growth than those in the other vessels (Figure 3). After 145 days, 98% of the protocorms in the flasks sealed with rubber bungs were above 0.5 mm in diameter, compared with 78% in tissue culture flasks. In contrast, no protocorms above 0.5 mm diameter were recorded in flasks equipped with breather tubes or with cotton wool bungs (Figure 3). Although the medium in the Petri dishes dried out in a relatively short period of time – these were discarded – there was no evidence of drying out in flasks equipped with breather tubes. Thus it seems that sealed containers promote growth rather than growth being depressed by drying out in flasks equipped with breather tubes.

When the sealed flasks were opened at the end of an experiment a distinct and characteristic unsaturated hydrocarbon smell was apparent. Thus a possible influence on the growth of the seedlings could have been the progressive accumulation of ethylene within the treatment flasks. Therefore, some flasks sealed with rubber bungs were sown with seed and some were left without seed, and at the end of a 145 day experimental period the concentrations of ethylene were determined. The levels of ethylene were markedly higher in flasks containing seedlings (125 μl dm^{-3}, compared with 17 μl dm^{-3} for control flasks) and the levels observed could perhaps be expected to produce a physiological response. However the low levels of ethylene which accumulated in control flasks suggested that a small amount of ethylene was being released from new rubber bungs, probably during autoclaving. New rubber bungs were therefore autoclaved twice before use.

In an attempt to investigate the importance of ethylene an experiment was performed in which small vials were included in the culture flasks containing potassium permanganate, as an absorbant of ethylene. Growth was clearly promoted in the presence of activated charcoal and permanganate alumina (Figure 4). It was also of

Figure 2. Apparatus used to supply specific gas mixtures to four culture flasks simultaneously.

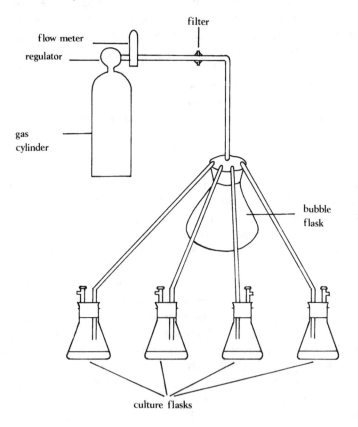

interest to note that the permanganate treatments caused a substantial reduction in the levels of ethylene in the flasks but that growth did not appear to be correlated with the accumulation of ethylene. Rather the concentration of carbon dioxide within the flasks appeared to exert a controlling influence.

In addition to changes in ethylene levels in closed containers it seemed probable that the developing protocorms and seedlings would cause changes in the concentrations of both carbon dioxide and oxygen within the flasks. To investigate this, seed was sown in sealed flasks and seedlings were grown under different photoperiodic conditions in an attempt to induce different seedling growth rates. Both protocorm growth and seedling fresh weight were enhanced by continuous illumination, while a 16 hour photoperiod was clearly less favourable (Table 1).

The concentration of oxygen and carbon dioxide were measured at intervals during the experiment. Figure 5 shows that growth in sealed containers resulted in marked progressive increases in the concentration of carbon dioxide within the flasks. However, a 16 hour day allowed the establishment of a much higher concentration of carbon dioxide than did continuous illumination. A 16 hour day also resulted in a much greater reduction in oxygen concentration than did continuous illumination which showed only a slight decline in the level of oxygen with time. It was therefore tempting to conclude that differences in growth could be explained, at least in part by differences in the concentration of carbon dioxide and oxygen within the flasks.

In order to examine these effects further flasks were flushed continuously with

Figure 3. The effect of different culture vessels on the diameter of protocorms of *Cattleya aurantiaca* after 145 days of growth under continuous illumination. Bars represent L.S.D. Culture vessels: ●, Flask with solid rubber bung; ○, 'Sterilin' tissue culture flask; ■, 'Sterilin' test tube; ❑, Flask with cotton wool bung; ▲, Flask with bung and breather tube; △, Boiling tube with cotton wool bung.

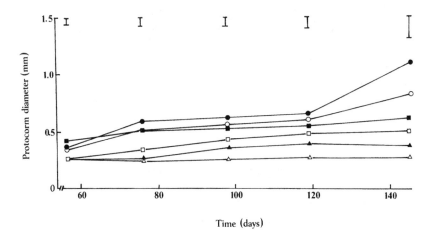

Time (days)

constant and predetermined gas mixtures containing specific concentrations of carbon dioxide and oxygen. Table 2 shows that with increasing carbon dioxide concentration germination was progressively inhibited when germination was recorded after 14 days, but this effect was overcome in part by a reduction in the concentration of

Table 1. *Effect of photoperiod on the development of seedlings. Data is mean ± S.E. of 100 seedlings withdrawn from each of 10 flasks.*

Photoperiod	Protocorm Development Index[a]	100 seedling fresh weight (g)[b]
Continuous illumination	215.4 ± 3.17	0.48 ± 0.11
16 h day	202.0 ± 1.23	0.16 ± 0.07

[a] after 77 days
[b] after 100 days

Figure 4. The effect of different absorbants on the fresh weight of seedlings of *Cattleya aurantiaca* after 100 days growth in relation to effects on the concentration of ethylene and carbon dioxide in the culture flasks. Bars represent S.E. A, Activated charcoal; B, Permanganate alumina; C, Permanganate vermiculite; D, Permanganate celite; E, Sealed flask; F, Flask with cotton wool bung.

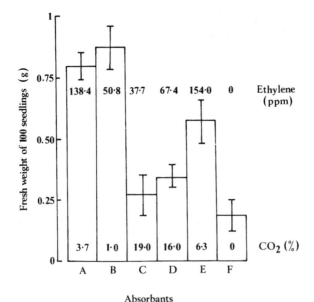

Table 2. *Effect of continuous flushing of flasks with different concentrations of carbon dioxide and oxygen on the germination (%) of* Cattleya aurantiaca *seeds under continuous illumination. 400 seeds scored for each treatment.*

Carbon dioxide (%)	Oxygen (%)	Days from start of germination		
		14	28	42
0.03	21	54	97	97
1.0	21	38	95	95
10	21	5	93	93
20	21	0	0	72
0.03	10	56	98	98
1.0	10	53	97	97
10	10	22	96	96
20	10	0	81	79
Sealed flask control		40	94	94

oxygen. After 28 days only the highest concentration of carbon dioxide was inhibitory and this effect was again less marked at 10% oxygen. Some of the initial differences observed in the level of germination were not reflected in effects on protocorm diameter after 50 days of growth. Thus, there was little difference in protocorm diameter in sealed flasks and flasks supplied with 0.03 and 1% carbon dioxide at 21% oxygen (Figure 6). Furthermore, reduced oxygen concentration now failed to promote growth at 0.03 and 1% carbon dioxide as measured by protocorm diameter. Higher concentrations of carbon dioxide were again inhibitory. At the end of the 100 day experimental period marked differences in the growth of seedlings were apparent. The fresh weight of seedlings in sealed flasks was similar to those in flasks flushed with 0.03% carbon dioxide and 20% oxygen (Figure 7), but an increase in carbon dioxide concentration to 1% resulted in a marked increase in fresh weight. It was of interest to note that growth was promoted further when both 0.03 and 1% carbon dioxide were supplied in the presence of 10% oxygen. Higher concentrations of carbon dioxide had inhibitory effects on growth possibly due to the delay in germination. Effects on seedling growth are also evident however at the lower carbon dioxide concentrations. As continual flushing of flasks was more favourable to growth than culture in a sealed container it was possible that some inhibitory factor was being removed.

Interestingly, the effects of the treatments imposed in this investigation on the dry weight of protocorms showed similar trends to those observed for fresh weights.

From a practical point of view carbon dioxide levels can be controlled by the inclusion of activated charcoal or alumina in the culture flasks or a simple and relatively inexpensive way to achieve additional growth and development was to vent sealed culture flasks at 30 day intervals. The protocorm development index in sealed containers was 241 and where flasks were flushed at 30 day intervals it was 387. As would be expected marked differences in the fresh and dry weights of these seedlings were also apparent (Figure 8).

Discussion

A wide variety of containers are employed by commercial growers and it seems to be a widely held view that it is important for culture vessels to allow a degree of gaseous exchange. In 1953, Breddy tested a number of different types of container and observed the best growth in test tubes with a solid bung fitted with a glass breather tube with cotton wool filter. Breddy suggested that growth was correlated with the degree of gas exchange and a tendency for the medium to dry out. Breddy did not include a sealed container in his trial, and Kano (1965) observed better growth of

Figure 5. Changes in the levels of oxygen (solid line) and carbon dioxide (dashed line) in sealed flasks during development of *Cattleya aurantiaca.* Seedlings under continuous illumination (▲) and 16 hour photoperiod (●). Bars represent L.S.D.

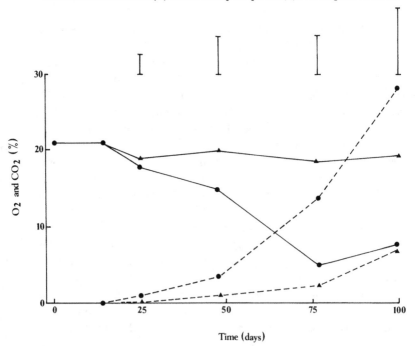

Time (days)

Figure 6. Effects of different concentrations of CO_2 and O_2 on the diameter of protocorms after 50 days growth under continuous illumination. Bars represent S.E.

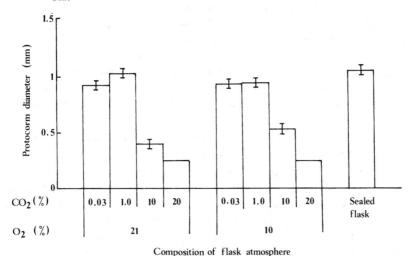

Figure 7. Effects of different concentrations of CO_2 and O_2 on the fresh weight of seedlings after 100 days growth under continuous illumination. Bars represent S.E.

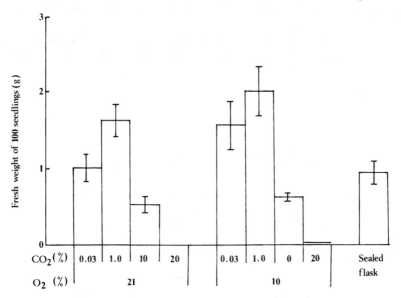

Laeliocattleya seedlings after 188 days in a sealed flask than in a flask containing a breather.

In this investigation when seedlings were grown in sealed containers there was no evidence of drying out of the medium and growth and development was markedly enhanced. One possible explanation involved the accumulation of ethylene in the culture vessels. Tamanaha *et al.* (1979) found that concentrations of ethephon up to 20 ppm (i.e. 20 mg dm^{-3}) slightly enhanced the development of leaves of seedlings of *Cattleya aurantiaca*, while concentrations above 50 ppm suppressed leaf development. The formation of roots was much more sensitive to ethylene in that inhibition was noted at concentrations above 2.5 ppm of ethephon. Although in this investigation ethylene accumulated in sealed flasks it seemed to be of less importance than either the concentration of carbon dioxide or oxygen in effects on growth and development. It was therefore apparent that growth could be promoted in sealed containers and the increase in carbon dioxide levels and the decrease in oxygen levels were of some significance. Where seeds were supplied with different concentrations of carbon dioxide and oxygen, high concentrations of carbon dioxide clearly delayed germination. Reducing the level of oxygen in the atmosphere overcame this effect in part at least. A similar ameliorating effect of reduced oxygen tension on germination at elevated CO_2 levels has been observed in *Galeola septentrionalis* (Nakamura *et al.* 1975).

Even though some seeds have very active fermentation pathways and the initial rapid production of ATP may be anaerobic (Pradet 1982), the germination of most

Figure 8. Effects of venting flasks at 30 day intervals on fresh (A) and dry weight (B) of 100 seedlings after 75 days growth under continuous illumination. Bars represent S.E.

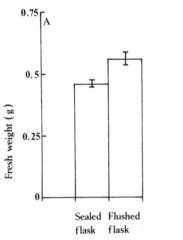
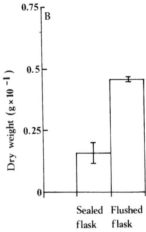

seeds is a process requiring oxygen. It is not uncommon for very high concentrations of carbon dioxide to inhibit germination. For example, Popay & Roberts (1970) reported that in *Capsella bursa-pastoris* and *Senecio vulgaris* germination could be inhibited by concentrations as low as 2% (i.e. 20 000 ppm).

It is perhaps worth noting that the substantial evolution of carbon dioxide observed with *Cattleya aurantiaca* in these experiments may not occur in other species. For example Yates & Curtis (1949) reported that when *Oncidium* was grown in sealed flasks, at different concentrations of sucrose, in the light or in the dark for more than one year, the concentrations of carbon dioxide ranged from 0.07 to 0.08%, and the oxygen concentration ranged from 20.39 to 21.29%. Thus they concluded that the differences in gas composition were inadequate to explain the observed differences in growth. However, Borg (1965) reported a 25% increase in growth with seedlings of a *Cymbidium* hybrid grown under Gro-lux lamps at a concentration of 2000 to 2300 ppm (0.2–0.23%) carbon dioxide.

McWilliams (1970) suggested that a large percentage of the more advanced epiphytic genera were CAM (Crassulacean Acid Metabolism) plants and Dressler & Dodson (1960) reported that the Epidendrae (which includes the genus *Cattleya*) may exhibit CAM to some extent. Furthermore, Knauft & Arditti (1969) detected labelled malate and citrate in leaves of *Cattleya*, which was interpreted in terms of a dark fixation pathway. In this investigation there was no evidence to support the view that protocorms and seedlings of *Cattleya aurantiaca* were exhibiting CAM behaviour. However, it may well be that *Cattleya aurantiaca* is a facultative CAM plant, using the C3 pathway (i.e. Calvin cycle) as seedlings but switching to the CAM pathway at some stage during development, or even when environmental conditions become adverse. The evidence in these experiments is consistent with the view that protocorms and seedlings of *Cattleya aurantiaca* exhibit the C3 system of carbon fixation, in that they respond in the expected way to changes in carbon dioxide and oxygen levels.

Much evidence exists to support the view that the rate of photosynthesis in C3 plants is usually limited by relatively low concentrations of carbon dioxide in the ambient atmosphere and it is further limited by the relatively high concentration of oxygen in the atmosphere (Tolbert & Zelitch 1983). It seems that in C3 plants oxygen competes with carbon dioxide in the regeneration of RuBP (Ribulose bisphosphate) reducing the efficiency of carboxylation. Furthermore the production of glycollate in the oxygenase reaction leads to the evolution of carbon dioxide in photorespiration (Pearcy & Bjorkman 1983). Thus part of the stimulation in growth when carbon dioxide concentrations are raised could be due to a reduction in oxygen inhibition due to competition of carbon dioxide for RuBP. High carbon dioxide and low oxygen favouring carboxylation of RuBP to phosphoglycerate thereby promoting carbon fixation. However this is only possible if the machinery is present and active in the protocorms. According to Harrison & Arditti (1978) the production of RuBP carboxylase is a major step towards autotrophy but this requires a source of carbon other than carbon dioxide. Indeed specific activity of RuBP reaches a maximum by day 30, a time which would almost coincide with the time of first leaf appearance.

Another factor which may be of some importance to the uptake of carbon dioxide is the resistances to flow encountered between the flask atmosphere and the sites of carboxylation in the chloroplast. There appears to be little or no information on the size of these resistances encountered in protocorms but it remains possible that they are substantial at least until leaf development has occurred around day 32-36.

Acknowledgements

We wish to thank Professor E.H. Roberts for his advice, interest and guidance.

References

Borg, F. (1965). Some experiments in growing *Cymbidium* seedlings. *Am. Orchid Soc. Bull.*, **34**, 899–902.

Breddy, N.C. (1953). Observations on the raising of orchids by asymbiotic cultures. *Am. Orchid Soc. Bull.*, **22**, 12–7.

Dressler, R.L. & Dodson, C.H. (1960). Classification and Phylogeny in the Orchidaceae. *Ann. Missouri Bot. Gard.*, **47**, 25–68.

Harrison, C.R. & Arditti, J. (1978). Physiological changes during the germination of *Cattleya aurantiaca* (Orchidaceae). *Bot. Gaz.*, **139**, 180–9.

Kano, K. (1965). Studies on the media for orchid seed germination. *Fac. Agric., Kagawa University*. No. 20, 1–68 (plus 6 plates).

Knauft, R.L. & Arditti, J. (1969). Partial identification of dark carbon dioxide fixation products in leaves of *Cattleya aurantiaca* (Orchidaceae). *New Phytol.*, **68**, 657–61.

McWilliams, E.L. (1970). Comparative rates of dark carbon dioxide uptake and acidification in the Bromeliaceae, Orchidaceae and Euphorbiaceae. *Bot. Gaz.*, **131**, 285–90.

Nakamura, S.J., Uchida, T. & Hamada, M. (1975). Atmospheric condition controlling the seed germination of an achlorophyllous orchid, *Galeola septentrionalis*. *Bot. Mag. Tokyo*, **88**, 103–9.

Pearcy, R.W. & Bjorkman, O. (1983). Physiological effects. In CO_2 and Plants, ed. E.R. Lemon, pp.65–98. *Am. Assoc. Adv. Sci. Selected Symposia*, **84**.

Popay, A.I. & Roberts, E.H. (1970). Factors involved in the dormancy and germination of *Capsella bursa-pastoris* (L.) and *Senecio vulgaris* (L.). *J. Ecol.*, **58**, 103–22.

Pradet, A. (1982). Oxidative phosphorylation in seeds during initial phases of germination. In *Physiology and Biochemistry of Seed Development, Dormancy and Germination*, ed. A.A. Khan, pp. 347–69. Amsterdam: Elsevier Biomedical Press.

Seaton, P.T. & Hailes, N.S.J. (1989). Effect of temperature and moisture content on the viability of *Cattleya aurantiaca* seed. In: *Modern Methods in Orchid Conservation: The Role of Physiology, Ecology and Management*, ed. H.W. Pritchard, pp. 17–29. Cambridge: Cambridge University Press.

Spoerl, E. (1948). Amino acids as sources of nitrogen for orchid embryos. *Am. J. Bot.*, **35**, 88–95.

Tamanaha, L.R., Craig, G.S. & Arditti, J. (1979). The effects of "Ethephon" on *Cattleya aurantiaca* (Orchidaceae) seedlings. *Bot. Gaz.*, **140**, 25–8.

Thompson, P.A. (1977). *Orchids from Seed*. Royal Botanic Gardens, Kew, Wakehurst Place: HMSO.

Tolbert, W.E. & Zelitch, I. (1983). Carbon Metabolism. In *CO_2 and Plants*, ed. E.R. Lemon, pp. 21–64. *Am. Assoc. Adv. Sci. Selected Symposia, 84,*.

Yates, R.C. & Curtis, J.T. (1949). The effect of sucrose and other factors on the shoot–root ratios of orchid seedlings. *Am. J. Bot.*, **36**, 390–6.

7

JOYCE STEWART

Orchid propagation by tissue culture techniques – past, present and future

Introduction

The discovery, about thirty years ago, that orchids could be asexually multiplied by a tissue culture technique (Morel 1960; 1964a,b) has led to an enormous increase in the number of plants, mostly artificial hybrids, in cultivation. Individual clones have been multiplied on a wide scale in many parts of the world either because their flower production can be controlled precisely, to meet heavy seasonal demand, or because of the quality, colour or longevity of their flowers. Since the first successful adaptation of *in vitro* techniques for the multiplication of *Cymbidium* clones, many other orchids have been investigated and many selected plants in more than 30 genera have been propagated in this way (Holdgate 1974; Murashige 1974; Morel 1974; Arditti 1977; Sagawa & Kunisaki 1982; George & Sherrington 1984) (Table 1). The reviews of Rao (1977) and Hughes (1981) list nearly twice as many genera but they included reports of orchids grown *in vitro* from seeds as well as tissue and organ cultures.

More recently it has been shown that, as for other plants, protoplasts can be isolated from the roots, stems, leaf tissue, petals and protocorms of orchids (Teo & Neumann 1978a,b; Pais *et al.* 1982; Price & Earle 1984; Loh & Rao 1985; Seeni & Abraham 1986) (Table 2). To date orchid protoplasts have been induced to grow and divide in culture but they have not stayed alive long enough to regenerate tissue or protocorms.

This paper surveys the work which has been reported on a wide range of orchids and notes features of the practice which will be of interest to those who wish to embark on this method of orchid propagation. It is primarily concerned with the regeneration of plantlets from organ and tissue culture and also includes a brief review of the progress with orchid protoplast culture.

The beginning

It was in 1955 that Morel (1960) cultured the minute apical meristem of virus-infected clones of *Cymbidium* in the hope of obtaining 'virus-free' plants, as had already been done with dahlia (Morel & Martin 1955), chrysanthemum (Holmes 1956) and carnation (Quak 1957). In this endeavour he was successful and he made three discoveries of far-reaching importance:

Table 1. *List of genera which have been propagated by tissue culture techniques as reported in the literature (There are probably other genera, especially hybrids, which are routinely multiplied in this way in commercial horticulture).*

European genera:		
Anacamptis	*Epipactis*	*Ophrys*
Dactylorhiza		

Tropical genera:		
Arundina	*Epidendrum*	*Pleione*
Cattleya alliance	*Laelia*	*Rhynchostylis*
Cymbidium	*Lycaste*	*Spathoglottis*
Cyrtopodium	*Oncidium* alliance	*Vanda*
Dendrobium	*Paphiopedilum*	*Vanilla*
Disa	*Phaius*	*Zygopetalum*
Doritis	*Phalaenopsis*	

Hybrid genera:		
Cattleya alliance	*Oncidium* alliance	*Vanda* alliance

Table 2. *List of orchid genera from which protoplasts have been isolated and cultured.*

European genera:		
Barlia		

Tropical genera:		
Acampe	*Cymbidium*	*Oberonia*
Aerides	*Dendrobium*	*Oncidium*
Angraecum	*Epidendrum*	*Paphiopedilum*
Brassavola	*Grammatophyllum*	*Rhynchostylis*
Brassia	*Geodorum*	*Spathoglottis*
Bulbophyllum	*Luisia*	*Vanda*
Calanthe	*Maxillaria*	*Vanilla*
Cattleya		

Hybrid genera:		
Aranda	*Oncidium*	*Renantanda*
Cattleya alliance	*Phalaenopsis*	*Vanda*
Dendrobium		

1. The cultured orchid meristem developed into a mass of tissue morphologically identical to the protocorm.
2. This protocorm sometimes developed a shoot and roots, but quite often it spontaneously produced a number of lateral protruberances or protocorm-like bodies.
3. The number of protocorms obtained from a single meristem could be enhanced artificially by cutting the first one into pieces and replacing the pieces on nutrient media, whereupon each piece proliferated to form several more protocorms which could be divided again or could be left to regenerate plantlets (Morel 1964b).

Meanwhile, it had been discovered that if the meristems were cultured in a liquid medium, and particularly if this was kept constantly in motion, the number of protocorm-like bodies could be greatly increased (Wimber 1963; 1965). Morel (1964b) estimated that, if frequent division of cymbidium protocorms could be continued successfully for one year, four million plantlets could be produced from a single apical meristem.

Within six years two papers revealed the rapid advances that had followed Morel's discovery. Bertsch (1966) reviewed the commercial significance of the new technique to the orchid industry. Vacherot (1966), representing one of the commercial firms most closely involved in developing the technique since its inception, hailed what he called 'meristem culture' of orchids as a 'breakthrough in the orchid world'. He divulged details of the experiments in his laboratories whereby 350 different clones of more than 10 genera were being multiplied in this way. The resulting plantlets are offered to orchid-growers as 'mericlones', a term coined by Crocker (Dillon 1964). Although they are said to have been derived from the parent plant by 'meristemming', the explant used in most commercial laboratories is often much larger than the apical meristem. It may consist of an entire lateral bud or some other organ (see below), and it would be more accurate, though less picturesque, to use the terms 'organ culture' or 'clonal propagation *in vitro*' for the provenance of these plantlets.

In any event, the 'mericlones' have proved to be mainly true to type, and, in the case of cymbidiums, usually 'virus-free'. Satisfactory techniques for the elimination of virus diseases from other genera have not been worked out yet. The primary objective in 'meristem culture' of orchids today is normally the rapid multiplication of awarded cultivars.

Review of some practical aspects of current techniques

Propagation material

The source of explants for tissue culture techniques among orchids was originally the apical meristem of a young shoot (Morel 1960). Terminal and axillary buds from shoot tips containing meristems have also been used on many occasions (Sagawa 1961; Morel 1963; 1964a,b; 1970; 1971; 1974; Vacherot 1966; Sagawa *et al.* 1966; Marston 1967; Reinert & Mohr 1967; Sagawa & Shoji 1967; Champagnat & Morel 1969; Lindemann *et al* . 1970; Kim *et al.* 1970; Vajrabhaya & Vajrabhaya 1970; Tse *et al.* 1971; Kunisaki *et al.* 1972; Fonnesbech 1972; Teo *et al.* 1973; Bubeck 1974; Intuwong & Sagawa 1974; Stewart & Button 1975; 1976b; Sagawa & Kunisaki 1982).

In those orchids where there is annual regrowth from underground tubers or rhizomes, such as the European ones, and all those which have a monopodial growth habit, the use of a shoot tip or bud can be very damaging. An important growing point of the plant must be removed. Frequently, in practice, it means that the whole plant is sacrificed. In researching a technique for a genus or species which has not yet been reported as amenable to tissue culture it is therefore necessary to have a number of plants available for experiment.

Trials have also been carried out with a number of other plant organs. Reports have been made of callus formation and plantlet regeneration from juvenile material such as protocorms (Pierik & Steegmans 1972) seedling leaf bases (Champagnat *et al.* 1970) and seedling leaf tips (Churchill *et al.* 1970). Young inflorescences or flower buds have been used for most of the vandaceous genera (Kotomori & Murashige 1965; Intuwong & Sagawa 1973; Sagawa & Kunisaki 1982; Len *et al.* 1986) and inflorescence primordia were used for the only reported successful multiplication of *Dactylorhiza* (as *Dactylorchis*) (Stokes 1974). Leaves have provided explant material in several instances (Churchill *et al.* 1973; Sagawa & Kunisaki 1982). Aerial roots have yielded callus and plantlets in *Epidendrum* (Stewart 1976; Stewart & Button 1978) and *Dendrobium* and *Phalaenopsis* (Sagawa & Kunisaki 1982).

Amongst the European orchids, the underground roots of *Neottia* have been shown to form protocorm-like bodies, buds and adventitious roots in culture (Champagnat 1971) as they do in nature. Pieces of the dormant, mature tuber of some species of *Ophrys* cultured *in vitro* have produced callus which has developed protocorms (Champagnat & Morel 1972; Morel 1974). The foliar embryos of *Hammarbya paludosa* (L.) O. Kuntze have not been propagated *in vitro*, so far as I am aware, but they are an important means of vegetative multiplication of populations in nature (Dickie 1875; Taylor 1967; Davies *et al.* 1983).

Decontamination of explants

Plant material in the greenhouse and garden is inevitably contaminated with a wide range of micro-organisms. Morel (1974) reported that 90% of cultures of *Paphiopedilum* stem apices are contaminated with bacteria. If underground organs are used as explants the contamination rate is even higher. Micro-organisms must be removed during the preparation of explants for *in vitro* culture because they would proliferate rapidly on the favourable substratum of a nutrient medium containing sucrose to the detriment of the explant and callus growth.

Several chemical agents are in common use for the surface-sterilization of plant material. The choice of agent and the time of treatment depend on the sensitivity of the material under investigation. Chlorine, released from freshly prepared sodium or calcium hypochlorite, seems to have been universally employed in decontaminating explants of orchids. It is claimed that its effectiveness is enhanced if a small amount of a fairly pure detergent, such as Teepol or Tween 20, is added to the sterilising solution as a wetting agent.

Details of sterilizing procedures which are widely used for a variety of plant materials have been summarised by Yeoman (1973). The pre-sterilization cleaning and post-sterilization rinsing to remove damaging chemicals are always important features of the routine. A technique worked out for orchids, in particular, is that of Sagawa *et al.* (1966) who advocate the sequential removal of enveloping leaves or bracts and immersion in gradually more dilute chlorine solutions until the explant is revealed and carefully rinsed before culturing. In many instances this lengthy treatment is unnecessary since the interior of the developing shoot or bud has not yet been exposed to micro-organisms and is therefore sterile.

With each new species or tissue investigated a routine procedure must be established. Often it is simply a tried technique with slight modifications to suit the material in hand.

Composition of nutrient media

Basal medium. 'Few plant tissues fail to respond to treatments designed to induce the formation of a callus and it now seems clear that the isolation and successful establishment of a callus largely depends on the culture conditions employed and not on the source of plant material' (Yeoman 1973). Foremost among the culture conditions is the discovery of a suitable nutrient medium which not only supports the life of the tissue but encourages its active growth and proliferation *in vitro.* The basis of all nutrient media is a source of carbon, usually sucrose, and a mixture of mineral salts combining the macro- and micro-nutrients in various concentrations.

Orchids which have responded to tissue culture methods of propagation can be divided into two groups, largely on the basis of their nutrient requirements *in vitro.* Many proliferate very easily on a very simple medium such as Knudson's C (1946), or the more stable Vacin & Went's medium (1949). Orchids in the *Cattleya* alliance, and others, have responded better to the somewhat more concentrated medium of Murashige & Skoog (1962). Many other media have been tried and a number have given good results in individual genera or species. These include those described by Heller (1953), Thomale (1954), Ojima & Fujiwara (1959), Morel (1965), Reinert &

Mohr (1967), Lindemann *et al.* (1970), and Schenk & Hildebrandt (1972), with slight modifications in individual cases. With each new investigation the optimal medium has to be found by experiment.

Additives . The response of isolated explants in culture depends on the endogenous growth substances at the time of excision and, in many cases, on the incorporation of various growth-regulating factors into the culture medium. Few excised tissues yield a useful callus, or multiple plantlets, without a supply of exogenous growth substances. Those which do proliferate include *Cymbidium*, and it is perhaps fortuitous that this was the first orchid genus to be cultured *in vitro*.

Most tissues can be placed in one of five groups depending on their requirements for growth substances in the culture medium.

1. Tissues which require an auxin in the culture medium.

    The beneficial effect of indole–3–acetic acid (IAA) or α–naphthalene acetic acid (NAA) on orchid seedlings has been known for a long time (Arditti 1967). One or other of these auxins at a concentration of 1 or 2 ppm (mg dm^{-3}) produces much higher survival rates for *Cattleya* meristems (Morel 1974) and they were also suggested for *Anacamptis* (Morel 1970). The synthetic auxin 2,4-dichlorophenoxyacetic acid (2,4-D) has been used successfully with *Paphiopedilum* (Morel 1974; Stewart & Button 1975) and with many genera when leaves or roots provide the explants (Sagawa & Kunisaki 1982).

2. Tissues which require a cytokinin in the culture medium.

    Amongst the orchids, even some of those which normally proliferate without any additive, it has been shown that a synthetic cytokinin, kinetin, has an enhancing effect (Fonnesbech 1972). For *Cattleya*, on the other hand, Reinert & Mohr (1967) showed that if kinetin was included in the culture medium from the beginning, growth of the explants was poor. In studies with *Cattleya* protocorms Pierik & Steegmans (1972) showed that the presence and concentration of benzylaminopurine (BA) affected development of the plantlets.

3. Tissues which require both an auxin and a cytokinin in the culture medium.

    Successful regeneration of protocorms from explants has been obtained in *Ophrys* when both NAA and kinetin were incorporated in the culture medium (Champagnat & Morel 1972; Morel 1974) and in *Paphiopedilum* when 2,4-D and BA were used (Stewart & Button 1975).

4. Tissues which require the addition of complex additives, usually coconut milk.

    There are many reports of orchid tissues which require the inclusion of fresh coconut milk, usually 10 to 15% (v/v) in the culture medium. These include explants from *Aranda* (Goh 1973), *Arundina* (Mitra 1971), *Ascofinetia* (Intuwong & Sagawa 1973), *Cattleya* (Scully 1967; Champagnat & Morel 1969); *Dendrobium* (Kim *et al* . 1970), *Paphiopedilum* (Bubeck 1974), *Phalaenopsis* (Kotomori & Murashige 1965; Intuwong & Sagawa 1974), and *Vanda* (Kunisaki *et al* . 1972; Teo *et al.* 1973).

5. Tissues which require an auxin and coconut milk.

Even more frequently coconut milk is used in addition to an auxin or auxin-like substance. The apparently synergistic effect of the two additives is very striking, not only in the increased proliferation of protocorms but in an increased survival rate for the original explants. Meristems excised from *Anacamptis* (Morel 1970), *Cattleya* (Scully 1967), *Dendrobium* (Sagawa & Shoji 1967), and *Rhynchostylis* (Vajrabhaya & Vajrabhaya 1970), among others, have produced plantlets when cultured *in vitro* on a medium containing both additives.

Other undefined additives which have been tried, but not so far recorded as essential, include peptone, casein hydrolysate, malt extract and yeast extract (Arditti 1977) and banana fruit (Sagawa & Kunisaki 1982).

Physical state of the nutrient medium. Two types of media have been used for the propagation of orchids by tissue culture techniques, namely solid and agitated liquid media. The solid, or semi-solid media are prepared by the addition of 6–10 g dm^{-3} Difco agar to the nutrient media and the explants are grown in flasks or tubes. Wimber (1963) introduced the use of a liquid medium, in flasks on a rotary shaker. He compared this with Morel's (1960) method using a solid medium, and with a stationary liquid medium, and found that the growth of the cymbidium explants ws 100% greater in the agitated liquid medium than in either of the others (Wimber 1965).

Two types of apparatus are used with liquid culture media. In the roller drum apparatus, which rotates at speeds of approximately 1 rpm, the tissue is alternately immersed in the nutrients and exposed to the air. Rotary shakers such as that originally used by Wimber (1963) are used at much greater speeds, usually 50–60 rpm, to ensure suspension of the tissue, which is completely immersed in the medium, and adequate aeration.

In many instances the choice between a liquid or solid medium rests on the availability of the apparatus or the whim of the investigator. It has been shown with *Cattleya*, however, that while it is possible to grow some cultivars on solid medium the results are much better with a liquid medium (Morel 1974) and for many varieties a liquid medium is essential for successful propagation (Scully 1967).

Since the result of a tissue culture investigation may depend on such a simple feature as the physical quality of the nutrient medium it would seem to be wise to try both solid and liquid types in any attempt to bring new species or varieties into axenic culture.

Changes of the culture medium during propagation. The establishment of an optimal series of sequential changes of the nutrient medium, both with regard to the presence, absence or concentration of growth hormones and to its physical quality, is one of the most important features of adapting tissue culture techniques for rapid clonal propagation. Murashige (1974) outlined the three stages of the tissue culture method, viz.

1. the establishment of an aseptic culture of the plant;
2. rapid multiplication of the propagula which will eventually give rise to new plants; and
3. preparation of these plantlets for their establishment in standard horticultural media.

With many orchids a single medium, such as Knudson C, has sufficed for all three stages. Several studies have shown that even better or swifter results can be attained by the use of a sequence of media (Lindemann *et al.* 1970; Intuwong & Sagawa 1974; Morel 1974). Such sequences usually involve an increase in the concentration of the growth hormones in the medium for stage (ii) as compared with stage (i) and a complete omission of auxin, cytokinin, or both, from the medium for stage (iii). The omission of sucrose from stage (ii) has also been beneficial in a range of genera (Sagawa & Kunisaki 1982). The incorporation of complex additives such as green banana fruit or pineapple juice for the final stage *in vitro* has been recommended (Morel 1974), as these substances promote rapid growth of orchid plantlets whether grown from asexually developed protocorms or from seeds.

Different physical types of media are also essential at the various stages of propagation. With *Cattleya* the 'starting medium' and 'multiplication medium' are both liquid while the final 'rooting medium' is solid (Lindemann *et al.* 1970).

Several reasons have been put forward to account for the different responses of cultures *in vitro* to the type of medium provided, and these have been reviewed by Murashige (1974). In practice, investigation of the most suitable medium for use at each stage of the tissue culture method must be made for each tissue of each new species or variety which it is intended should be propagated in this way.

Morphogenesis of plantlets

In the *in vitro* propagation of orchids, several pathways have been observed in the morphogenesis of plantlets.

Dormant bud development. The simplest is the development of axillary buds on cultured node sections such as those from flower stalks of *Dendrobium* (Singh & Sagawa 1972), *Epidendrum* (Stewart & Button 1976a), *Phalaenopsis* (Kotomori & Murashige 1965) and others. The optimal conditions of temperature and humidity provided by the growth room and the culture flask together with the release from apical dominance encourages the development of every bud whose potential is only occasionally realised under natural conditions.

Protocorm development. The most frequent mode of development is through a series of events peculiar to the orchid family. Diffuse growth of the explant occurs by random cell division and enlargement and after some weeks it develops into a proliferative body (or group of bodies) morphologically identical to the protocorm. This term was first used by Bernard (1904) to describe a stage of development in the orchid embryo. The adventive protocorms which proliferate on cultured shoot tips and other explants develop into small plantlets in exactly the same way as those developed from seed embryos. As noted above, they can be artificially divided into several pieces and proliferation occurs from the epidermis of each piece, four to five new protocorms being found around its periphery. In *Dactylorhiza* and a number of epiphytic orchids this pathway is followed (Stokes 1974; Arditti 1977).

Several anatomical studies have shown that the origin of the adventive protocorms varies. They arise from the epidermis of the explant in *Cymbidium* (Champagnat *et al.* 1966; Rutkowski 1971), from meristematic cells near to the vascular system at the base of bud scales in *Cattleya* (Champagnat & Morel 1969) and from the hypodermis in the young inflorescence explants of the intergeneric hybrid *Ascofinetia* (Intuwong & Sagawa 1973). Individually separated cells grew, multiplied and became organised into protocorms in a study undertaken by Steward & Mapes (1971).

Callus development. Sometimes the original explant first develops an undifferentiated callus mass on the surface of which numerous small protocorms will appear. Each of these protocorms can be removed for propagation and will eventually form a plantlet. This type of regeneration has been described in the *Odontoglossum* alliance (Morel 1971; 1974), in several of the vandaceous orchids including *Aranda* (Goh 1973) and *Rhynchostylis* (Vajrabhaya & Vajrabhaya 1970), in *Ophrys* (Champagnat & Morel 1972; Morel 1974) and in *Paphiopedilum* (Morel 1974; Stewart & Button 1975; 1976b).

It is worth noting that similar callus structures occasionally develop from germinating orchid seeds. They have been known for many years in *Vanda* (Curtis & Nichol 1948; Rao 1963) and I have observed them on germinating seeds of *Paphiopedilum spicerianum* and *Epidendrum* hybrids (Stewart 1976).

Root and bud development. A fourth type of morphogenesis has been described by Bubeck (1974) and Stewart (1976) for *Paphiopedilum*, in which roots and buds differentiated from vascular bundles in the cultured apical meristem. Both these organs subsequently developed into plantlets and up to 10 plantlets could be obtained from each cultured meristem.

Protoplast culture

Protoplasts have been isolated from orchid protocorms (Teo & Neumann 1978a), leaves (Teo & Neumann 1978b; Pais *et al.* 1982; Price & Earle 1984; Loh & Rao 1985; Seeni & Abraham 1986), root tips (Seeni & Abraham 1986) and flowers (Price & Earle 1984). There appear to be no difficulties in isolating ample quantities of clean and metabolically active protoplasts from different sources of tissue in more than 20 wild species and hybrids. The standard techniques such as those reviewed by Bajaj (1977) and Cocking & Evans (1977) have been followed.

Protoplasts have been cultured in Knudson C (Teo & Neumann 1978a) and in Vacin & Went media (Loh & Rao 1985; Seeni & Abraham 1986) with the additives 2,4-D, NAA, kinetin, BA and coconut milk in various combinations. After 10 days in culture 10–13% of the protoplasts in media with growth regulators were dividing. Small clusters of cells were observed after about 20 days in culture. The cells lysed at this stage which suggests that the culture medium may be unsuitable for their continued development. No callus tissue or protocorms were observed in these studies. However, cymbidium plants were obtained from free cell cultures by Steward & Mapes (1971) which makes it seem likely that with refinements of the method adult plants of other orchids could be regenerated from protoplasts. In view of the great variations in media which are currently used for the multiplication of orchids in tissue culture it seems likely that the conditions necessary for the growth of cell clusters in protoplast culture will need extensive investigation.

Nevertheless, the isolation and culture of orchid protoplasts has been described as one of the most significant developments in the field of plant tissue culture in recent years (Bajaj 1977). The techniques have far-reaching implications for plant improvement by cell manipulation and somatic hybridization if cell fusion can be effected. Rao (1977) has suggested that protoplast culture techniques may become useful in the production of new varieties or hybrids of orchids, particularly where normal hybridization and embryo culture techniques have failed.

I would like to suggest that such techniques could also provide invaluable methods for conservation if clonal material can be regenerated. The possibility of the propagation of rare and endangered species from one or a few plants can be envisaged. One gram of plant tissue yields a very large number of protoplasts. This could be obtained from part of a leaf, thus obviating the need to destroy shoots on rare plants in cultivation or even to remove a plant from the wild. However, the experimental work must be carried out first, utilising species which are readily available in quantity.

The future

It will be apparent that among those species and hybrids which have been tried a very large number can be propagated by a tissue culture technique. But it must be admitted that only a fraction of the most easily cultivated orchids has so far been investigated. Some entire groups with a rather special morphology, such as the terrestrial orchids with tubers or tuberoids or fibrous roots have scarcely been examined. Most work has concentrated on the rapid multiplication of desirable clones for horticulture, chiefly hybrids. Many species await investigation. The work is tedious and time-consuming.

Protocorms and plantlets developed in sterile culture have provided useful and easily handled research material and it is expected that more details of seedling physiology will become available from their continued use.

So far as conservation is concerned the method has not yet been utilised, so far as I am aware. Our recent attempts to use it for the endangered *Cypripedium calceolus* in Britain (Stewart & Marlow unpublished) have been hampered by the paucity of material available for experiment. When a species has become reduced in numbers to a very few plants it may already be too late to save it: methods need to be worked out before a species becomes endangered. The use of protoplast cultures, if they are successfully developed in future, may be useful as a means of clonal propagation for rarities as well as in the field of somatic hybridization.

References
Arditti, J. (1967). Factors affecting the germination of orchid seeds. *Bot. Rev.*, 33, 1–97.
Arditti, J. (1977). Clonal propagation of orchids by means of tissue culture – a manual. In *Orchid Biology Reviews and Perspectives I*, ed. J. Arditti, pp. 203–93. Ithaca and London: Cornell University Press.
Bajaj, Y.P.S. (1977). Protoplast isolation, culture and somatic hybridization. In *Applied and Fundamental Aspects of Plant Cell, Tissue and Organ Culture*, ed. J. Reinert and Y.P.S. Bajaj, pp. 467–96. Berlin: Springer–Verlag.
Bernard, N. (1904). Recherches experimentales sur les orchidées. *Revue Gén. Bot.*, 16, 405–51; 458–78.
Bertsch, W. (1966). A new frontier: orchid propagation by meristem tissue culture. In *Proc. Fifth World Orchid Conf.*, ed. L.R. de Garmo, pp. 225–9. Long Beach, California: Fifth World Orchid Conference Inc.
Bubeck, S.K. (1974). A study of Paphiopedilum meristem culture. Ph.D. Thesis (unpublished). New Brunswick: Rutgers University.
Champagnat, M. (1971). Recherches sur la multiplication végétative de *Neottia nidus-avis* Rich. *Ann. Sci. Nat. Bot., Sér. 12*, 12, 209–47.
Champagnat, M. & Morel, G. (1969). Multiplication végétative des *Cattleya* a partir de bourgeons cultivés *in vitro*. *Soc. Bot. Fr. Mém.*, 116, 111–32.
Champagnat, M. & Morel, G.M. (1972). La culture in vitro des tissus de tubercules d'*Ophrys*. *C.R. Hebd. Séanc. Acad. Sci., Paris*, 274, 3379–80.
Champagnat, M., Morel, G., Chabut, P. & Cognet, A.M. (1966). Recherches morphologiques et histologiques sur la multiplication végétative de quelques orchidées du genre *Cymbidium*. *Revue Gén. Bot.*, 73, 706–46.
Champagnat, M., Morel, G. & Mounetou, B. (1970). La multiplication végétative des *Cattleya* a partir de jeunes feuilles cultivées aseptiquement *in vitro*. *Ann. Sci. Nat. Bot., Sér. 12*, 11, 97–114.
Churchill, M.E., Ball, E.A. & Arditti, J. (1970). Production of orchid plants from seedling leaf tips. *Orchid Digest*, 34, 271–3.

Churchill, M.E., Ball, E.A. & Arditti, J. (1973). Tissue culture of orchids I. Methods for leaf tips. *New Phytol.*, **72**, 161–6.

Cocking, E.C. & Evans, P.K. (1977). The isolation of protoplasts. In *Plant Tissue and Cell Culture*, ed. H.E. Street, pp.103–35. Oxford: Blackwell Scientific Publications.

Curtis, J.T. & Nichol, M.A. (1948). Culture of proliferating orchid embryos *in vitro*. *Bull. Torrey Bot. Club.*, **75**, 358–73.

Davies, P., Davies, J. & Huxley, A. (1983). *Wild Orchids of Britain and Europe*. London: Chatto and Windus, The Hogarth Press.

Dickie, G. (1875). Note on the buds developed on leaves of *Malaxis*. *J. Linn. Soc. (Bot.).*, **14**, 1–3.

Dillon, G.W. (1964). The meristem Merry-Go-Round. *Am. Orchid Soc. Bull.*, **33**, 1023–4.

Fonnesbech, M. (1972). Growth hormones and propagation of *Cymbidium in vitro*. *Physiol. Plant.*, **27**, 310–16.

George, E.F. & Sherrington, P.D. (1984). *Plant Propagation by Tissue Culture*. Eversley, Basingstoke: Exegetics Ltd.

Goh, C.J. (1973). Meristem culture of *Aranda* Deborah. *Malayan Orchid Rev.*, **12**, 9–12.

Heller, R. (1953). Recherches sur la nutrition minérale des tissus végétaux cultivées *in vitro*. *Ann. Sci. Nat. Bot.*, **14**, 1–223.

Holdgate, D.P. (1974). The tissue culture of orchids. *Orchid Rev.*, **82**, 58–61.

Holmes, F.O. (1956). Elimination of aspermy virus from the Nightingale chrysanthemum. *Phytopathology*, **46**, 599–600.

Hughes, K.W. (1981). Ornamental species. In *Cloning Agricultural Plants via* in vitro *Techniques*, ed. B.V. Conger, pp. 5–50. Boca Raton, Florida: CRC Press Inc.

Intuwong, O. & Sagawa, Y. (1973). Clonal propagation of Sarcanthine orchids by aseptic culture of inflorescences. *Am. Orchid Soc. Bull.*, **42**, 209–15.

Intuwong, O. & Sagawa, Y. (1974). Clonal propagation of *Phalaenopsis* by shoot tip culture. *Am. Orchid Soc. Bull.*, **43**, 893–5.

Kim, K.K., Kunisaki, J.T. & Sagawa, Y. (1970). Shoot-tip culture of Dendrobiums. *Am. Orchid Soc. Bull.*, **39**, 1077–80.

Knudson, L. (1946). A new nutrient solution for the germination of orchid seeds. *Am. Orchid Soc. Bull.*, **15**, 214–17.

Kotomori, S. & Murashige, T. (1965). Some aspects of aseptic propagation of orchids. *Am. Orchid Soc. Bull.*, **34**, 484–9.

Kunisaki, J.T., Kim, K.K. & Sagawa, Y. (1972). Shoot tip culture of *Vanda*. *Am. Orchid Soc. Bull.*, **41**, 435–9.

Len, L.-H.C., Choon, T.-L.G. & Kheng, P.L. (1986). Clonal propagation of orchids from flower buds. In *Proc. Fifth ASEAN Orchid Congress Seminar* , ed. A.N. Rao, pp. 98–101. Singapore.

Lindemann, E.G.P., Gunckel, J.E. & Davidson, O.W. (1970). Meristem culture of *Cattleya*. *Am. Orchid Soc. Bull.*, **39**, 1002–4.

Loh, C.S. & Rao, A.N. (1985). Isolation and culture of mesophyll protoplasts of *Aranda* Noorah Alsagoff. *Malayan Orchid Rev.*, **19**, 34–7.

Marston, M.E. (1967). Clonal multiplication of orchids by shoot meristem culture. *Scient. Hort.*, **19**, 80–6.

Mitra, G.C. (1971). Studies on seeds, shoot tips and stem discs of an orchid grown in aseptic culture. *Indian J. Exp. Bot.*, **9**, 79–85.

Morel, G.M. (1960). Producing virus-free cymbidiums. *Am. Orchid Soc. Bull.*, **29**, 495–7.

Morel, G.M. (1963). La culture in vitro du méristème de certaines Orchidées. *C.R. Hebd. Séanc. Acad. Sci., Paris*, **256**, 4955–7.

Morel, G.M. (1964a). La culture *in vitro* du méristème apical. *Revue Cytol. Biol. Vég.*, *Paris*, **27**, 307–14.

Morel, G.M. (1964b). Tissue culture – a new means of clonal propagation of orchids. *Am. Orchid Soc. Bull.*, **33**, 473–8.

Morel, G.M. (1965). Eine neue Methode erbgleicher Vermekrung: die Kultur von Triebspitzen-Meristemen. *Die Orchidee*, **16**, 165–76.

Morel, G.M. (1970). Neues auf dem Gebiet der Meristem-Forschung. *Die Orchidee*, **20**, 433–43.

Morel, G.M. (1971). The principles of clonal propagation of orchids. In *Proc. Sixth World Orchid Conf.*, ed. M.J.G. Corrigan, pp. 101–6. Sydney: Sixth World Orchid Conference.

Morel, G.M. (1974). Clonal multiplication of orchids. In *The Orchids: Scientific Studies*, ed. C.L. Withner, pp. 169–222. New York: John Wiley & Sons.

Morel, G.M. & Martin, C. (1955). Guérison des plantes atteintes de maladies à virus; par culture de méristèmes apicaux. In *Report of the XIV International Horticultural Congress*, The Hague-Scheveningen, ed. J.P. Nieuwstraten, pp. 303–10. Wageningen: H. Veenman & Zonen.

Murashige, T. (1974). Plant propagation through tissue cultures. *Ann. Rev. Pl. Physiol.*, **25**, 135–66.

Murashige, T. & Skoog, F. (1962). A revised medium for rapid growth and bioassays with tobacco tissue cultures. *Physiol. Plant.* **15**, 472–97.

Ojima, K. & Fujiwara, A. (1959). Studies on the growth promoting substance of the excised wheat roots. I. Effects of peptone on the growth. *Tohoku J. Agricu. Res.*, **10**, 111–28.

Pais, M.S.S., Anjos, F. & Rangel de Lima, M.A. (1982). Obtention of protoplasts from the terrestrial orchid *Barlia longibracteata*, experimental conditions and inframicroscopy. In *Plant Tissue and Cell Culture 1982*, ed. A. Fujiwara, pp. 599–600. Tokyo: The Japanese Association for Plant Tissue Culture.

Pierik, R.L.M. & Steegmans, H.H.M. (1972). The effect of 6–benzylamino purine on growth and development of *Cattleya* seedlings grown from unripe seeds. *Z. Pflanzen Physiol.*, **68**, 228–34.

Price, G.E. & Earle, E. (1984). Sources of orchid protoplasts for fusion experiments. *Am. Orchid Soc. Bull.*, **53**, 1035–43.

Quak, F. (1957). Meristeemcultur, gecombineerd met warmtebehandeling, voor het verkrijgen van virus-vrije anjerplanten. *Tijdschr. Pl. Ziekt.*, **63**, 13–14.

Rao, A.N. (1963). Organogenesis in callus cultures of orchid seeds. In *Plant Tissue and Organ Culture, a Symposium*, ed. P. Maheshwari & N.S. Ranga Swamy, pp. 332–43. Delhi: International Society of Plant Morphologists.

Rao, A.N. (1977). Tissue culture in the orchid industry. In *Applied and Fundamental Aspects of Plant Cell, Tissue and Organ Culture*, ed. J. Reinert & Y.P.S. Bajaj, pp. 44–69. Berlin: Springer–Verlag.

Reinert, R.A. & Mohr, H.C. (1967). Propagation of *Cattleya* by tissue cultures of lateral bud meristems. *Proc. Am. Soc. Hortic. Sci.*, **91**, 664–71.

Rutkowski, E. (1971). How meristems multiply. *Am. Orchid Soc. Bull.*, **40**, 616–22.

Sagawa, Y. (1961). Vegetative propagation of phalaenopsis stem cuttings. *Am. Orchid Soc. Bull.*, **30**, 808–9.

Sagawa, Y. & Kunisaki, J.T. (1982). Clonal propagation of orchids by tissue culture. In: *Plant Tissue and Cell Culture 1982*, ed. A. Fujiwara, pp. 683–4. Tokyo: The Japanese Association for Plant Tissue Culture.

Sagawa, Y., Shoji, T. & Shoji, T. (1966). Clonal propagation of cymbidiums through shoot meristem culture. *Am. Orchid Soc. Bull.*, **35**, 118–22.

Sagawa, Y. & Shoji, T. (1967). Clonal propagation of dendrobiums through shoot meristem culture. *Am. Orchid Soc. Bull.*, **36**, 856–9.

Schenk, R.U. & Hildebrandt, A.C. (1972). Medium and techniques for induction and growth of monocotyledonous and dicotyledonous plant cell cultures. *Can. J. Bot.*, **50**, 199–204.

Scully, R.M. (1967). Aspects of meristem culture in the *Cattleya* alliance. *Am. Orchid Soc. Bull.*, **36**, 103–8.

Seeni, S. & Abraham, A. (1986). Screening of wild species and hybrids of orchids for protoplast isolation. In *Proc. Fifth ASEAN Orchid Congress Seminar*, ed. A.N. Rao, pp. 23–7. Singapore: Parks & Recreation Dept., Ministry of National Development.

Singh, H. & Sagawa, Y. (1972). Vegetative propagation of *Dendrobium* by flower stalk cuttings. *Hawaii Orchid J.*, **1**, 19.

Steward, F.C. & Mapes, M.O. (1971). Morphogenesis in aseptic cell cultures of *Cymbidium. Bot. Gaz.*, **132**, 65–70.

Stewart, J. (1976). Clonal propagation of *Paphiopedilum* and *Epidendrum* with particular reference to tissue culture techniques. M.Sc. Thesis (unpublished). Pietermaritzburg: University of Natal.

Stewart, J. & Button, J. (1975). Tissue culture studies in *Paphiopedilum. Am. Orchid Soc. Bull.*, **44**, 591–9.

Stewart, J. & Button, J. (1976a). Rapid vegetative multiplication of *Epidendrum* × *O'Brienianum in vitro* and in the greenhouse. *Am. Orchid Soc. Bull.*, **45**, 922–30.

Stewart, J. & Button, J. (1976b). Tissue culture studies in *Paphiopedilum*. In *Proc. Eighth World Orchid Conf.*, ed. K. Senghas, pp. 372–8. Frankfurt am Main: German Orchid Society Inc.

Stewart, J. & Button, J. (1978). Development of callus and plantlets from *Epidendrum* root tips cultured *in vitro. Am. Orchid Soc. Bull.*, **47**, 607–12.

Stokes, M.J. (1974). The *in vitro* propagation of *Dactylorchis fuchsii* (Druce) Vermeul. *Orchid Rev.*, **82**, 62–5.

Taylor, R.L. (1967). The foliar embryos of *Malaxis paludosa. Can. J. Bot.*, **45**, 1553–6.

Teo, C.K.H., Kunisaki, J.T. & Sagawa, Y. (1973). Clonal propagation of strap-leafed *Vanda* by shoot-tip culture. *Am. Orchid Soc. Bull.*, **42**, 402–5.

Teo, C.K.H. & Neumann, K.H. (1978a). The culture of protoplasts isolated from *Renantanda* Rosalind Cheok. *Orchid Rev.*, **86**, 156–8.

Teo, C.K.H. & Neumann, K.H. (1978b). The isolation and hybridization of protoplasts from orchids. *Orchid Rev.*, **86**, 186–9.

Thomale, H. (1954). *Die Orchideen.* Stuttgart: Verlag Eugen Ulmer.

Tse, A.T.-Y., Smith, R.J. & Hackett, W.P. (1971). Adventitious shoot formation on *Phalaenopsis* nodes. *Am. Orchid Soc. Bull.*, **40**, 807–10.

Vacherot, M. (1966). Meristem tissue culture propagation of orchids. In *Proc. Fifth World Orchid Conf.*, ed. L.R. De Garmo, pp. 23–6. Long Beach, California: Fifth World Orchid Conference Inc.

Vacin, E.F. & Went, F.W. (1949). Some pH changes in nutrient solutions. *Bot. Gaz.*, **110**, 605–13.

Vajrabhaya, M. & Vajrabhaya, T. (1970). Tissue culture of *Rhynchostylis gigantea*, a monopodial orchid. *Am. Orchid Soc. Bull.*, **39**, 907–10.

Wimber, D.E. (1963). Clonal multiplication of cymbidiums through tissue culture of the shoot meristem. *Am. Orchid Soc. Bull.*, **31**, 117–20.

Wimber, D.E. (1965). Additional observations on clonal multiplication of cymbidiums through culture of shoot meristem. *Cymbidium Soc. News*, **20**, 7–10.

Yeoman, M.M. (1973). Tissue (callus) Cultures – Techniques. In: *Plant Tissue and Cell Culture*, ed. H.E. Street, pp. 31–58. Oxford: Blackwell Scientific Publications.

8

MICHAEL J. HUTCHINGS

Population biology and conservation of *Ophrys sphegodes*

Introduction

The Red Data Book for vascular plants describes the present status in the British Isles of the early spider orchid, *Ophrys sphegodes* Mill., as 'vulnerable' (Perring & Farrell 1983). In the last half-century, its distribution has declined in extent to the point that it now occurs in only about 20% as many 10×10 km squares as it had been recorded in prior to 1930 (Figure 1), and many of the remaining populations in this country consist of very few plants. If the rate of decline observed over the last fifty years were to continue, the species would disappear from the British Isles by the end of this century.

Sensible arguments can be made both for and against the likelihood of this dramatic rate of decline being maintained. However, even given the most favourable prognosis, the species is clearly under serious threat. The rapid decline of the range of *O. sphegodes* towards the south-eastern corner of the British Isles suggests that a climatic deterioration might be a factor which has been involved in the reduction in its

Figure 1. Distribution maps of *Ophrys sphegodes* in the British Isles. A, Pre-1930, B, Since 1975.

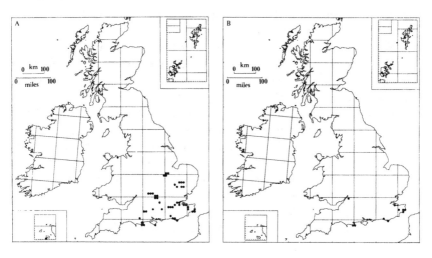

geographical distribution. Apart from this, the most obvious extrinsic factors involved in its decline appear to be habitat interference and habitat destruction. In addition, however, there are intrinsic life historical characteristics which may help to explain why *O. sphegodes* has rapidly become less abundant. The identification of such characteristics is an important step which will help us to develop management schemes for its conservation. However, these characteristics should not be regarded as necessarily unique to *O. sphegodes*; it is possible that the rarity of other species of orchid has similar causes.Thus, a consideration of these characteristics may be a useful contribution both to the study of orchid ecology and to orchid conservation.

In the first part of this paper I review features of the life history of *O. sphegodes* which appear to have contributed to it becoming a threatened species; some of these characteristics are found throughout the Orchidaceae. In the second, I discuss those features of the behaviour of *O. sphegodes* which lead to the conclusion that it is, among the Orchidaceae native to the British Isles, a relatively weedy species. This fact is likely to influence the way in which its habitats are managed for conservation. Thirdly, a summary of the influence of recent management upon the status of a population of *O. sphegodes* is presented. Finally, a plea is made for the adoption of those methods to monitor the effects of management upon rare species which will allow maximum interpretation of the demographic status of the species. Throughout the following discussion, data concerning the population biology of *O. sphegodes* were obtained during a long-term demographic study of the largest remaining population in the British Isles (Hutchings 1987a,b).

Aspects of the life history of *Ophrys sphegodes*

Pollination biology

Different species of bees and wasps are reported as displaying a high fidelity in their pollination of species within the genus *Ophrys* (Tutin *et al.* 1980). However, the literature contains little information about the particular species which are the most common pollinators of *O. sphegodes*. Clapham et al. (1962) state that the species is visited sparingly by bees, and that seed is set only if the plants are visited, thus implying that self-pollination is not possible, although field observations suggest that it can occur. Nilsson (1979) states that pollination is usually by bees of the genus *Andrena*. The question of how a pollinating species with a high fidelity or flower-constancy comes to pollinate such a rare resource as *O. sphegodes*, remains to be resolved. It may be that a pheromone is produced that attracts male bees, as has been found in certain other orchid species (Fahn 1979). However, the poor rate of seed set in this country appears to indicate that lack of pollinators is a major limitation to seed production. Whereas it has been reported that nearly half of the capsules in some populations of *O. sphegodes* from the Continent contain ripe seeds, the corresponding figure for populations in this country is only about 10% (personal observation; Lang

1980). Maturation of *O. sphegodes* seeds may take up to 3 months, and although seed viability may be relatively good (i.e. 50–60% germination, see Muir, Chapter 4 this Volume) the proportion of full seeds produced, i.e. containing an embryo, is low – perhaps no more than 44% (H.W. Pritchard, pers. comm.).

Post-germination life-history: the subterranean phase

In those species, such as members of the family Orchidaceae, which have very small ('dust') seeds, some form of anomalous nutrition is found, involving reliance on obtaining resources for growth by such means as symbiosis, saprophytism or parasitism (Harper *et al.* 1970; Hadley & Pegg, Chapter 5 this Volume). Such forms of nutrition are necessary because most of the weight of seeds, apart from the embryo, consists of food reserves which are drawn upon to fund growth immediately after germination. These reserves are reduced to very small amounts in small-seeded species. In the Orchidaceae, seed weight is of the order of 10^{-5}–10^{-6} g (Harper *et al.* 1970). The post-germination development of orchid protocorms and mycorrhizomes is a slow process, largely because of their small initial size and consequent inability to fund the production of leaves and flowers with their limited resources. No above-ground growth occurs in many species of orchid for several years after germination (Wells 1981). Many estimations of the length of the subterranean phase of life in species of orchid appear to have been as much the result of guesswork as of careful recording. Despite observations of the study population of *O. sphegodes* for twelve years to date, the length of the subterranean phase of development in this species is still unknown.

Post-germination life-history: the aerial phase

Even after entering a phase of life during which aerial structures can be produced, many orchid species spend several years in a vegetative state before flowering for the first time (see Wells 1981; and Table 1). During the prolonged period of time which can pass between seed dispersal and flowering, each plant may fall victim to any one of a wide variety of mortality risks, including drought, waterlogging, mechanical damage and predation. During this period there is no potential for leaving sexually-produced offspring, although some daughter plants may arise by vegetative propagation. Since vegetative propagation is an activity with a high energetic demand, the production of daughter plants by this means is unlikely to be rapid. In the types of habitat which *O. sphegodes* occupies, namely ancient, species-rich chalk and limestone grassland, where competition for scarce resources is often severe, vegetative propagation is a characteristic seen in many species. It enables species to avoid complete dependence on the high risk seed regeneration phase (Grubb 1977) and extends life-spans of established plants. However, *O. sphegodes* does not possess efficient vegetative propagation; less than 5% of the emergent population appears to be recruited in this way each year (Hutchings 1987a).

Table 1. *Summary table of known life-history characteristics for species of the Orchidaceae. Maximum recorded ages may reflect the length of the studies rather than the maximum life-spans of the species concerned.*

Species	Half-life (years)[a]	Time from emergence to flowering (years)	Percentage of emergent population flowering each year	Maximum recorded age (years)	Source
Dactylorchis sambucina	19.2 (dry wooded meadow) 20.8 (mesic wooded meadow)	7–14	Range 0–81.6%, mean 25.5% (dry wooded meadow). Range 16.1–69.2%, mean 40.8% (mesic wooded meadow). Overall mean (14 years mean) = 30.5%	30	Tamm (1972)
Dactylorchis incarnata	—	4–5	0–27.8%	25	Tamm (1972)
Listera ovata	83.6	0–9	Range 3.1–86.2% Overall mean (21 years) = 46.0%	28	Tamm (1972)
Orchis mascula	4.3	6–8	Range 0–100% Mean (14 years) = 25.0%	14	Tamm (1972)
Orchis simia	—	3+	—	10	Willems (1982)
Orchis militaris	10.25 (Buckinghamshire) 5.96 (Suffolk) (these are depletion rather than survivorship curves)	—	Overall mean for 18 years at two sites = 26.7%	10+	Farrell (1985)

Table 1 (contd.)

Species	Half-life (years)[a]	Time from emergence to flowering (years)	Percentage of emergent population flowering each year	Maximum recorded age (years)	Source
Aceras anthropophorum	4.0–7.8	—	Overall mean (14 years) = 41.1%	14	Wells (1981)
Spiranthes spiralis	4.6–9.2	13–15	Overall mean (14 years) = 32.8%	—	Wells (1981)
Herminium monorchis	0.5–6.6 yrs (mean 3.8+)	—	Overall mean (14 years) = 16.9%	—	Wells (1981 and pers. comm.)
Gymnadenia conopsea	—	—	Overall mean (10 years) = 47.3%	—	Hutchings (unpublished)
Ophrys apifera	6.4–13.4	0–6	—	—	Wells (pers. comm.)
Ophrys sphegodes	1.5–2.3	0–4	Range 65–97%. Overall mean (10 years) = 84.0%	10	Hutchings (1987a)

[a]Half-lives are based on survivorship curves in all cases except *Orchis militaris*, and exclude the subterranean phase of life.

The period from first emergence above ground until flowering is particularly short in *O. sphegodes*, compared with many other orchid species for which data have been collected (Table 1). Indeed, over 70% of plants of *O. sphegodes* flower in the first year they emerge above ground, and 100% have flowered at least once by their fourth year above ground. To some extent, this capacity of surviving plants to enter a phase of sexual maturity early in the above ground phase of life would offset the adverse effects on population growth of a lengthy pre-emergence period. However, detailed analysis of data collected for a number of species of the Orchidaceae from temperate habitats now indicates that the lives of individual plants which have reached the aerial phase of growth are punctuated by irregularly-spaced periods of dormancy (Wells 1967; Tamm 1972; Hutchings 1987a). Such periods are commonly of up to two years duration in different species. Detailed analysis of the behaviour of *O. sphegodes* indicates that in every year approximately 50% of the population is in a dormant state, and that there is a >95% probability that plants which have not been recorded above ground for three consecutive years are dead (Hutchings 1987a).

Several subtle age-related changes occur in the behaviour of plants of *O. sphegodes* which have reached the aerial phase of life (Hutchings 1987b). Mention has already been made of the increase in probability of flowering with length of time since first emergence. Performance, as measured by the mean height of flower spikes, increases from the first year of emergence above ground for between 4 to 7 years and then declines again (Figure 2A). In addition, several behavioural characteristics are related to life span. The probability of emerging above ground in any year declines significantly with increasing life-span (Figure 2B). The probability of flowering in any year also declines with increasing life-span, although this fall just fails to achieve significance (Figure 2C). Conversely, there is a significant increase, with increasing life-span, in the probability of emergence being accompanied by flowering; plants which live for longer emerge at wider intervals of time, but are more likely to flower when they do emerge than are shorter-lived plants (Figure 2D). This observation gives rise to the speculation that flowering is a debilitating activity, and that it is the cause of the dormant periods, and a contributory cause of death, in *O. sphegodes*. However, analysis of the available evidence does not support this hypothesis. The probability of flowering is significantly greater for orchids which flowered in the previous year than for orchids which were either vegetative or dormant in the previous year (Figure 3A). Nevertheless, even orchids which flower in any year have a probability of only 0.43 of flowering in the next year. Dormancy in fact turns out, in *O. sphegodes*, to be the surest route to an early grave, since a dormant plant has a probability of 0.80 of also being dormant in the succeeding year (Figure 3B). Regardless of behaviour in any given year, *O. sphegodes* has a high probability of dying in the next year, or at least of dying without emerging again; even for those plants which flower in any year, there is a probability of 0.33 that they will die without ever being recorded again (Figure 3C).

It is becoming clear from this analysis of the life history of *O. sphegodes* that it is beset by high mortality risks at many points throughout its life, and thus life-span will be short. In fact, the age-specific mortality risk (Begon & Mortimer 1986; Varley & Gradwell 1970) for this species amounts to more than a 50% probability of death in the twelve months after its first appearance above ground (Figure 4). For the small number of recruited plants that have survived for one year after the first production of

Figure 2. Relationships between measures of performance of *Ophrys sphegodes* plants and (in A) length of time in years since plants first emerged above ground and, (in B–D), life span of orchid. A, Mean (± S.E.) flower spike height (cm). The fitted quadratic is significant at p<0.05. B, Ratio (mean number of appearances above ground/life span) plotted against life span. The slope of the line is significant (*p*<0.001, $t_{[8]}$ = -8.25). C, Ratio (mean number of flowering episodes/life span) plotted against life span. The slope of the line just fails to reach significance (*p*<0.06, $t_{[8]}$ = -2.27). D, Ratio (mean number of flowering episodes/mean number of appearances above ground) plotted against life span. The slope of the line is significant (p<0.01, $t_{[8]}$ = 4.68).

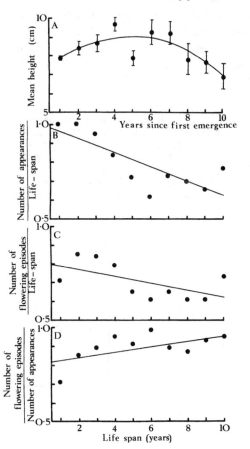

leaves, there is a declining mortality risk as plants age further. However, beyond the sixth year of life, the age-specific mortality risk increases rapidly with age, which indicates that maximum life-span is being approached. A parameter used by plant demographers to describe the life-spans of plants in populations is the half-life. This is exactly analogous to the half-life used to characterise the rate of decay of a radioisotope, and it measures the length of time that it would take for half of a population of plants, all recruited within a particular interval of time, to die. Ranges of half-lives which have been calculated for several different orchid species are given in Table 1. The half-life for *O. sphegodes* is particularly short; more than half of the new plants recruited in a given year would be dead within about two years of first emerging above ground. This fact underlines the problem which is created for this species when there is a failure to recruit offspring through sexual reproduction. Unless regular recruitment is achieved *via* seeds, a population of *O. sphegodes* will rapidly decrease in numbers towards extinction, since vegetative propagation is so limited as a means of population maintenance.

An additional hazard faced by *O. sphegodes* is herbivory. Although the species does not appear to suffer from serious invertebrate grazing, it is heavily grazed by sheep, as witnessed by the removal of nearly all flower spikes in a year in which grazing was allowed on the site supporting the population. Only small flower spikes were missed by the grazers (Hutchings 1987a).

Figure 3. Mean probabilities that *Ophrys sphegodes* will display various types of behaviour in year $n+1$ after either flowering (F), producing a vegetative rosette (V), or remaining dormant (D) in year n. A, Probability of flowering in year $n+1$. B, Probability of being dormant in year $n+1$. C, Probability of dying without reappearing after year n. Within each graph, significant differences ($p<0.05$) in the probabilities of types of behaviour being displayed are denoted by different letters of the alphabet. In each graph, some plants placed in the category 'dormant' may already be dead.

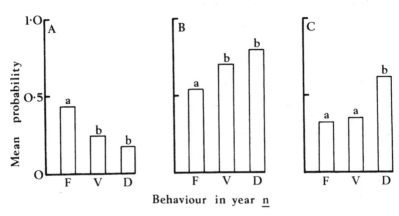

Behaviour in year n

Ophrys sphegodes as a weedy species

At least three life historical characteristics of *O. sphegodes* make it appear, relative to other species in the Orchidaceae, to be an r-selected species (i.e. the population is on the rising sector of an asymptotic growth curve, and mortality is mainly independent of density) rather than a K-selected species (i.e. the population is near asymptotic density and mortality is largely density dependent) (Table 1):

1. it has a very short half-life
2. it achieves sexual maturity quickly after entering the aerial phase of life.
3. a very high proportion of emergent plants flower. Over ten years of study a mean of 84% of those plants with above ground parts during the flowering season produced flower spikes. This final characteristic needs to be

Figure 4. Age-specific mortality rate (q_x) and killing power (k_x) (see Begon & Mortimer 1986; and Varley & Gradwell 1970), determined in different years after first emergence of *Ophrys sphegodes*.

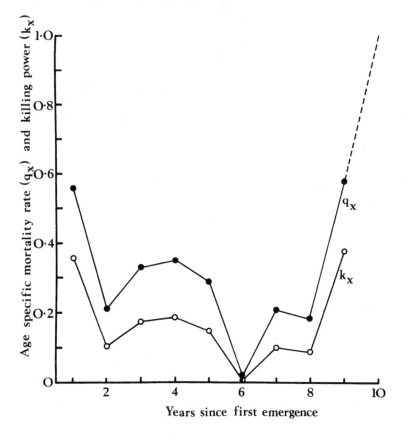

carefully interpreted. If data had been collected when rosettes had first emerged above ground, in the previous October, it is possible that this proportion, based on emergent plants, would be substantially lower, bringing it more into line with values for other orchid species.

On the basis of this evidence, *O. sphegodes* appears to be the most weedy species among the Orchidaceae for which accurate data have yet been assembled. The intuitive

Figure 5. Contribution of *Ophrys sphegodes* plants of different ages to a measure of the net reproductive rate (R'O) of the population. A, Absolute contribution to net reproductive rate. B, Proportional contribution to net reproductive rate.

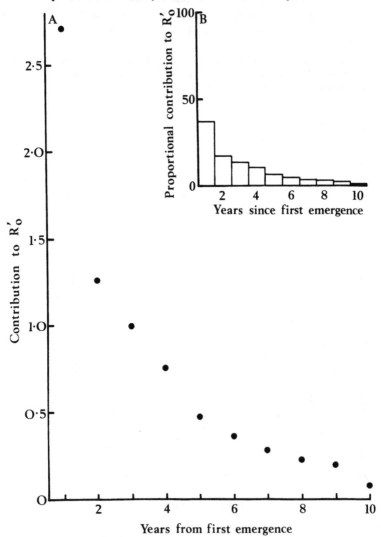

view that members of the Orchidaceae are long lived species of persistently stable habitats must be revised in the light of such evidence. It might be predicted that this species would display a tendency towards rapid invasion when conditions allow. Field observations made in 1986 showed that the population upon which this study is based began to invade adjacent abandoned arable land, thus supporting this suggestion. The tendency to display invasions of new sites is also a well-known feature of other orchid species including *O. apifera*, which raises the question of whether the genus *Ophrys* is a relatively r-selected genus within the Orchidaceae.

Unlike many weedy species, *O. sphegodes* is not strictly monocarpic. Nevertheless, although many plants in this study are known to have flowered on more than one occasion, there is a considerable majority that only flowered once. This fact, combined with the high mortality risks experienced by plants in their first year above ground, results in the major flowering effort for the population under study being invested in the youngest plants. On average, in any year, plants which have emerged above ground for the first time produce 37% of the flowers, and plants in their first two years above ground produce 54% of flowers (Figure 5). The main contribution to future recruitment into the population, given that this is dominated by seed recruitment, will clearly be made by young plants; thus, maintenance of conditions favourable to flowering, seed maturation and seedling establishment is an essential part of any management scheme designed for the continued conservation of *O. sphegodes*.

The effects of recent management upon the status of the population of *Ophrys sphegodes*

The form of management at present in use on the site of the study population involves light grazing by sheep throughout the year except for the period from May to August, when the population is flowering and setting seed. Grazing helps to maintain a short sward in which plants of low stature and low competitiveness, like *O. sphegodes*, can survive. The success of the current management regime can be judged from Figure 6, which illustrates in full the extent of the flux which has taken place in the population since recording began in 1975. Before 1980, the site was winter-grazed by cattle, which caused much trampling damage to the vegetation and soil surface, and probably caused severe mechanical damage to the parts of plants below ground. From 1980 onwards the site has been grazed by sheep. In 1980, sheep were allowed to graze the site continuously, with the result that nearly all flower spikes were eaten (Figure 7; and Hutchings 1987a). Although no quantitative measures are available for seed output, recruitment of new plants from seed would have been dangerously low in that year because of poor seed production. Figure 6 illustrates declining recruitment and a high rate of loss of plants from the population until 1980, when the management regime was altered, since which recruitment has been consistently high, and far in excess of mortality. The net effect is clear. Until 1980, the population was in rapid decline, displaying a cumulative loss, over the six years of the study until that date, of 97 plants. From 1981 until 1984 there was a cumulative net gain of 129 plants, resulting in a net gain into the population over the ten year period of the study, of 32 plants. Although data for 1985 and 1986 have not been analysed in detail yet, this improved trend has been maintained, and this was reflected in a record above ground population in 1985.

Population censusing as an aid in the management of rare species

Plant population biology has been a fashionable field of research for the last twenty years. In the main, however, it has remained an academic pursuit, with little of

Figure 6. Analysis of flux in a population of *Ophrys sphegodes* studied from 1975–1984. Cumulative recruitment (▲), annual recruitment (△), cumulative mortality (■), annual mortality (□), cumulative net yearly change (●), annual net yearly change (○).

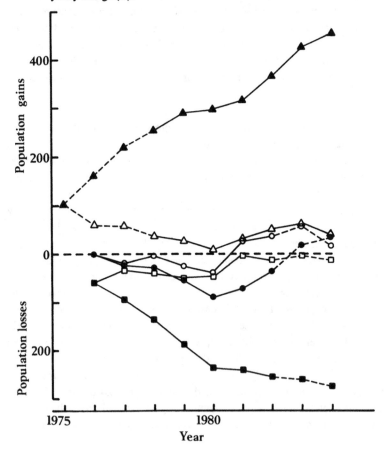

its knowledge or methodology being transmitted to the practices of those involved with species management. It seems appropriate to use this opportunity to make a plea for the adoption of census techniques, rather than mere counts, for assessing the effects of management regimes and changes in management regimes upon rare species populations. Unfortunately, census techniques are laborious, time-consuming and physically wearing to undertake. Nevertheless, they provide the only means, apart from undesirable destructive techniques, by which the *flux* in a population – that is to say the numbers of recruitments (births) and deaths over given periods of time – and the age-structure of a population, can be assessed. This type of information can not be obtained by counting. An illustration of this is provided by the presentation of actual counts of the emergent population of *O. sphegodes* from 1975–1984 (Figure 7). This graph gives cause for concern because of the low number of plants in the middle years of the study, and shows an apparently satisfactory recovery in numbers. However, it gives no hint of the change in the condition of the population before and after 1980. Without the year-by-year information concerning numbers of births and deaths which could be gleaned from repeated censuses, the downward trend in the population might not have been rectified until it was too late to save the population from extinction. The use of census techniques has a long and distinguished history in a number of situations where management of vegetation has been monitored (White 1985), yet even now there are numerous situations where it is applicable but not utilised.

Figure 7. Number of plants observed (●), number flowering (■), number grazed (▲), and estimated total size (○) of a population of *Ophrys sphegodes* from 1975–1984.

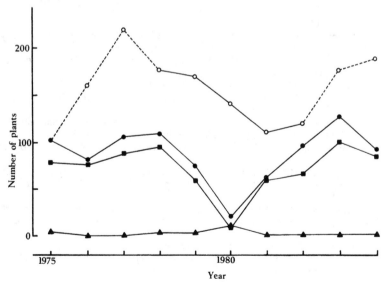

In addition to this there is the question of the time of year when censuses should be carried out. There is an instinctive tendency to manage attractive species to maximize their apparency during the most attractive phase of their annual cycles; in the case of the Orchidaceae, this is the flowering phase. However, as in the case of *O. sphegodes*, a high proportion of the population may be dormant or vegetative each year. It may be more instructive for determining population size, to conduct censuses outside the flowering season, for example just after emergence of vegetative rosettes each year. In the case of *O. sphegodes* this would be in October. Our opinion of the status of rare species might undergo dramatic revision if this approach is utilised. For some species (although probably not *O. sphegodes*), conservation in a vegetative state may be just as valuable, at least in the short term, as maintenance of sexually reproducing populations, in just the same way that preservation of dormant, viable seeds in the soil may be a natural way of conserving species which have been thought extinct.

Acknowledgements

I am grateful to the Nature Conservancy Council for providing access to the study site, and for helpful discussion of all aspects of the work, and to C.D. Preston of the Biological Records Centre, Institute of Terrestrial Ecology, for research into the past and present distribution of *O. sphegodes*, and for production of the maps. Many people assisted with the collection of data; without their vital contribution this study would have been impossible. I thank A. Maclellan for his critical reading of earlier drafts of the manuscript.

References

Begon, M. & Mortimer, M. (1986). *Population Ecology: A Unified Study of Animals and Plants*. Oxford: Blackwell Scientific Publications.

Clapham, A.R., Tutin, T.G. & Warburg, E.F. (1962). *Flora of the British Isles*. Cambridge: Cambridge University Press.

Fahn, A. (1979). *Secretory Tissues in Plants*. London: Academic Press.

Farrell, L. (1985). Biological Flora of the British Isles, No.160. *Orchis militaris* L. *J. Ecol.*, **73**, 1041–53.

Grubb, P.J. (1977). The maintenance of species-richness in plant communities: the importance of the regeneration niche. *Biol. Rev.*, **52**, 107–45.

Harper, J.L., Lovell, P.H. & Moore, K.G. (1970). The shapes and sizes of seeds. *Ann. Rev. Ecol. Syst.*, **1**, 327–56.

Hutchings, M.J. (1987a). The population biology of the early spider orchid, *Ophrys sphegodes* Mill. 1. A demographic study from 1975–1984. *J. Ecol.*, **75**, 711–27.

Hutchings, M.J. (1987b). The population biology of the early spider orchid, *Ophrys sphegodes* Mill. 2. Temporal patterns in behaviour. *J. Ecol.*, **75**, 729–42.

Lang, D. (1980). *Orchids of Britain*. Oxford: Oxford University Press.

Nilsson, S. (1979). *Orchids of Northern Europe*. Harmondsworth, Middlesex: Penguin Books Ltd.

Perring, F.H. & Farrell, L. (1983). *British Red Data Book: 1. Vascular Plants.* Lincoln: Society for the Promotion of Nature Conservation.

Tamm, C.O. (1972). Survival and flowering of some perennial herbs. 2. The behaviour of some orchids on permanent plots. *Oikos,* 23, 23–8.

Tutin, T.G., Heywood, V.H., Burges, N.A., Moore, D.M., Valentine, D.H., Walters, S.M. & Webb, D.A. (eds.). (1980). *Flora Europaea, Volume 5,* Alismataceae to Orchidaceae (Monocotyledons). Cambridge: Cambridge University Press.

Varley, G.C. & Gradwell, G.R. (1970). Recent advances in insect population dynamics. *Ann. Rev. Entomol.,* 15, 1–24.

Wells, T.C.E. (1967). Changes in a population of *Spiranthes spiralis* (L.) Chevall. at Knocking Hoe National Nature Reserve, Bedfordshire, 1962-65. *J. Ecol.,* 55, 83–99.

Wells, T.C.E. (1981). Population ecology of terrestrial orchids. In *The Biological Aspects of Rare Plant Conservation,* ed. H. Synge, pp. 281–95. Chichester: Wiley.

White, J. (1985). The census of plants in vegetation. In *The Population Structure of Vegetation, Handbook of Vegetation Science 3,* ed. J. White, pp. 33–88. Dordrecht: Dr. W. Junk.

Willems, J.H. (1982). Establishment and development of a population of *Orchis simia* Lamk. in the Netherlands, 1972-1981. *New Phytol.,* 91, 757–65.

Predicting population trends in *Ophrys sphegodes* Mill.

Introduction

The successful conservation of threatened species necessitates the protection and management of their habitats. If species abundance is low, i.e. the species under consideration is locally rare, management regimes must be developed and applied with extreme care if local extinction is to be avoided. The vulnerability of such populations to interference will severely restrict the extent to which experimental management trials can be used to develop measures to ensure conservation. Under these circumstances, population modelling may provide a valuable alternative to field trials. In this paper, the development and use of a simple model of the population behaviour of the early spider orchid, *Ophrys sphegodes* Mill. is described, and used to predict future population trends.

O. sphegodes is a rare species which occurs in ancient, species-rich chalk grassland. Its distribution and abundance in the British Isles have declined considerably over the past 50 years. It is considered to be part of the European element of the British flora (Summerhayes 1951), and its distribution is currently virtually confined to the South East corner of England. Rosettes appear above ground in autumn, persist over winter and may produce flower spikes in April or May in the following year. Shortly after flowering the above ground parts senesce (Lang 1980; Hutchings 1987a). Most plants have short life-spans, and reach sexual maturity rapidly. Few plants survive for more than three years after their first emergence above ground (Hutchings 1987a; and Hutchings, chapter 8 this Volume).

Methods

The demography of *O. sphegodes* has been studied on a south-west facing slope in chalk grassland at Castle Hill, National Nature Reserve (NNR), Sussex. The population is the largest remaining in Great Britain. All orchids occurring in a 20 by 20 m permanent square plot have been censused every May since 1975, during the flowering period. The x and y coordinates and status (i.e. flowering or vegetative) of each individual is recorded, together with various measures of performance. Plant coordinates allow scaled maps to be prepared, showing the location of individual plants within the plot. The life history of individual orchids may be determined by comparing maps obtained from successive years. Plants may enter a dormant state,

Table 1. *Mean plant state transition probabilities, and mortality rates of vegetative, flowering and dormant plants of O. sphegodes. Values calculated on data collected between 1975 and 1985.*

Behaviour in year n + 1	Behaviour in year n		
	Flowering	Vegetative	Dormant
flowering	0.43	0.24	0.19
vegetative	0.02	0.05	0.01
dormant	0.54	0.70	0.80
Mortality rate of:	0.33	0.35	0.61

which normally lasts for no more than two years, after which they may reappear in the emergent population. Thus, although dormant plants are not recorded during mapping, they are not classified as dead. During the ten years of this study from 1975 to 1984 the mean density of *O. sphegodes* plants was 0.32 m^{-2}. Further details of the site and of the population behaviour of *O. sphegodes* may be found in Hutchings (1983; 1987a,b; and Chapter 8 this Volume).

The model

The population of orchids recorded each year may be divided into plants in three states – flowering individuals, vegetative individuals and dormant individuals. Dormant plants were further classified as either first year or second year dormant plants. From a knowledge of the fate of individual plants, it is possible to calculate the average probability that a plant present in a given state in one year will survive to the next year in the same state or in different state (Hutchings 1987a). These probabilities are presented in Table 1. A transition matrix T (Begon & Mortimer 1986), which describes the behaviour of the population may be derived directly from the values in Table 1. For example, the probability that a dormant plant survives to the next year is equal to 1-0.61, and the probability that it will be a vegetative above ground plant is 0.01. Thus the combined probability that a dormant plant will survive one year and appear the next year as a vegetative rosette, is given by; $(1-0.61) \times 0.01 = 0.0039$ (Figure 1). Individual elements of T give the probability that a plant of a given state will survive to the next year, appearing either in the same state or in a changed state. The diagonal elements of the matrix are the probabilities that individuals survive unchanged from one year to the next. Thus the first value in the first row (t(1,1)), gives the probability that a vegetative plant will survive to the following year, and appear again above ground as a vegetative individual (i.e. 3.27% of vegetative plants appear the following year as vegetative individuals, Figure 1C). Off-diagonal elements give the probabilities that individuals will appear the following year in a changed state. The first element of the second row (t(2,1)) is 0.157, thus on average 15.7% of vegetative plants appear the following year as flowering plants (Figure 1).

The composition of the population present at any time may be represented by a column vector X (Figure 1B). Each element of X denotes the number of plants present in the population in a particular state. Thus x(1) is the number of vegetative plants present, x(2) the number of flowering plants, while x(3) and x(4) are the numbers of first and second year dormant plants respectively. The number of plants surviving and the composition of the population in the absence of recruitment may be obtained by multiplying the transition matrix T by the column vector, i.e. TX(t) = X(t+1). Each

Figure 1. A, Basic transition model for *O. sphegodes* population. B, T = transition matrix, X = column vector, where X_1 represents the number of vegetative plants, X_2 the number of flowering plants, and X_3 and X_4 the numbers of first and second year dormant plants respectively. C, Individual t values gives the probability that a plant present one year in a given state will survive to the following in the same or a changed state.

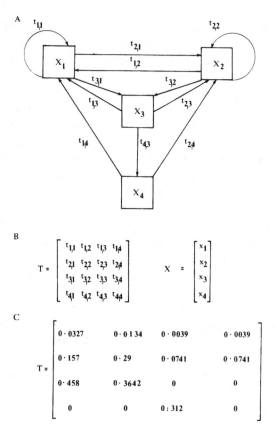

Table 2. *Calculated reproductive values (R), of*
flowering Ophrys sphegodes *plants. R equals the*
number of recruits recorded in year t divided by the
number of flowering plants present in year t – n.

n	Mean	Lowest value	Highest value
1	0.606	0.27	0.92
2	0.577	0.28	1.02
3	0.486	0.31	0.84
4	0.550	0.33	1.03
5	0.500	0.39	0.64
6	0.503	0.31	0.68
7	0.587	0.33	0.80
8	0.580	0.39	0.77

successive multiplication will produce a new column vector $X(t+n)$, the sum of whose elements is equal to the total population size. The transition matrix remains unchanged. In the absence of density dependent population growth, recruitment may be considered to be a direct function of the number of flowering plants; the number of new plants entering the population in a given year will be a function of the number of plants which flowered at some date n years earlier. Flowering plants may be ascribed a reproductive value R, where R is equal to the number of recruits recorded in year t, divided by the number of plants which flowered in year t–n. The effective reproductive values of flowering plants remains approximately constant regardless of the length of time delay (n years), used in their calculation (Table 2). Mean values of R range from 0.486 to 0.606 recruits per flowering plant as the time (n years) from seed production to above ground recruitment is altered. The lack of significant variation in R enables recruitment to be modelled easily without the need to incorporate a time delay factor between seed and emergence. The total number of recruits in year t+1 is obtained by multiplying the number of flowering plants present in year t by R. Over the study period, 30% of recorded recruits were vegetative during their first year above ground. The remaining 70% flowered. Hence the numbers of new vegetative plants is equal to 0.3 × (number of flowering plants × R), while the number of new flowering individuals is given by 0.7 × (number of flowering plants × R).

It is impossible to determine from the available field data, whether a significant number of recruits enter the growing population as dormant individuals. However, using the model, the effects of dormant plant recruitment may be examined by assuming that a fixed proportion of the total above ground recruits enter the first year dormant plant class each year.

The values of the elements in the final transition matrix M, on which the model output is based, depend on R, the reproductive value of flowering plants, P, the proportion of above ground recruits entering the population as dormant individuals, and the overall levels of plant mortality.

Matrix M =

0.0327	0.0134 + 0.3R	0.0039	0.0039
0.1570	0.2900 + 0.7R	0.0741	0.0741
0.4580	0.3642 + RP	0	0
0	0	0.312	0

The behaviour of this type of matrix may be characterised by two parameters, the dominant eigenvalue (λ) and the corresponding eigenvector (Searle 1966). The eigenvalue represents the rate of population growth from one year to the next. For a stable population $\lambda = 1.0$; if λ is less than 1.0 the population decreases, if λ is greater than 1.0 the population increases. The eigenvector describes the proportional distribution of the population between the four states (flowering, vegetative, first and second year dormant). The level of mortality that a population is able to sustain without decreasing in size may be estimated using the formula: $H = 100((\lambda-1)/\lambda)$, where H is the percentage of the population which may be lost from each class of the population (Jeffers 1978).

Matrix elements were altered in a systematic manner. The behaviour of each resulting matrix was determined by calculating its dominant eigenvalue and eigenvector using the method of power (direct) iteration (Burden *et al.* 1981). This approach allows assessment of the sensitivity of the population to changes in mortality and recruitment rates.

Results

Recruitment

In the absence of recruitment the population declines rapidly. Figure 2 illustrates the pattern of decline which the population of *O. sphegodes* present in 1985 would demonstrate if recruitment of new plants was to cease. Within three years vegetative plants are lost from the population and the number of flowering plants decreases from 115 to 11 plants.

Plant mortality

The model output suggests that populations of *O. sphegodes* are particularly sensitive to increased mortality of flowering plants (Figure 3). An increase in flowering plant mortality rate of 5% from the mean recorded value of 33% is sufficient to cause the population to decline. A similar absolute increase in the mortality rate of vegetative plants would also cause the population to decline i.e. $\lambda < 1.0$ (Figure 3). Conversely, a positive rate of population growth ($\lambda > 1.0$), is maintained as dormant plant mortality is increased from the recorded mean value of 61% to 70% (Figure 3). The graph of the relationships between λ and plant mortality rate shows that population growth in *O. sphegodes* is most sensitive to changes in flowering plant mortality rate, and least sensitive to changes in vegetative plant mortality rate (Figure 3).

Reproductive values and dormant plant recruitment
Population growth rates (λ) increase linearly as the reproductive value of

Figure 2. Predicted decline of the 1985 population of *O. sphegodes* at Castle Hill, in the absence of recruitment. N is the total population size, D is the number of dormant plants, F is the number of flowering plants and V is the number of vegetative plants.

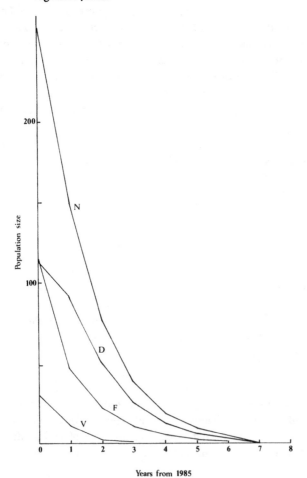

Years from 1985

flowering plants (R) is increased (Figure 4). The sensitivity of the population to increased levels of recruitment is strongly influenced by the proportion (P) of recruits entering the population as dormant individuals. When P is zero, no recruitment of dormant individuals occurs and an R value greater than 0.878 is required simply to maintain the population. As P increases the critical value of R required to sustain the population decreases rapidly. Observed yearly rates of proportionate population increase ranged from 0.79 to 1.55, whilst R values varied from approximately 0.3 to 1.0 (Table 2). Given the restricted range of recorded R values, substantial levels of dormant plant recruitment must occur if the observed rates of population increase are to be achieved. The model suggests that approximately four times as many dormant recruits as above ground recruits are required to enter the population, in order to achieve this, i.e. P = 4 (Figure 4). In practice, P, like the other model parameters would not be a constant.

Figure 3. The effects of dormant plant mortality (D), vegetative plant mortality (V) and flowering plant mortality (F) on the predicted growth rate (λ) of *O. sphegodes* population.

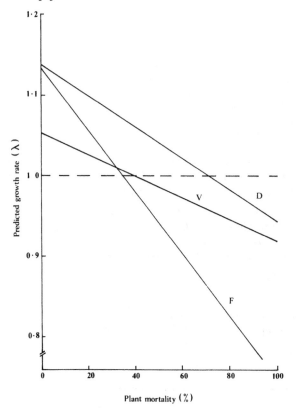

Figure 4. The effects of dormant plant recruitment levels (P), and flowering plant reproductive value (R) on the predicted growth rate (λ) of *O. sphegodes* populations.

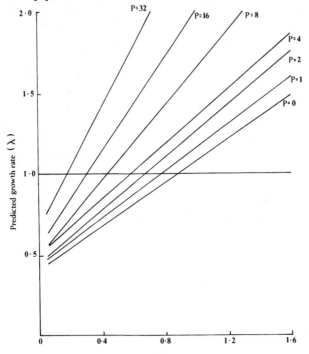

R (reproductive value of flowering plant)

Table 3. *Critical flowering plant reproductive values, required for the population to maintain a constant size i.e.* $\lambda = 1.0$; R_c, *critical reproductive value; P, level of dormant plant recruitment; H, average percentage of population which may be harvested, based on mean of recorded R values greater than R_c.*

P	R_c	% frequency R_c exceeded in any year	H
4.0	0.581	57.1	14.4
2.0	0.681	57.1	9.6
1.0	0.776	28.5	6.3
0	0.878	14.3	2.7

When rates of plant mortality are held constant the minimum reproductive value (R) required to maintain the population is governed by the levels of dormant plant recruitment (Table 3). If four times as many above ground plants as dormant plants enter the population (i.e. P = 4), a minimum flowering plant reproductive value of 0.581 is required. This level of reproductive performance may be expected to occur only approximately 6 times in any decade (Table 3). As levels of dormant plant recruitment are reduced, the critical value of R required to maintain the population increases, and the likelihood of this value being exceeded in any year decreases, along with the additional level of plant mortality (H), that the population is capable of withstanding. Actual calculated flowering plant reproductive values are frequently below those required to maintain the population (Table 2).

Discussion

In the absence of recruitment the population of *O. sphegodes* at Castle Hill will decline rapidly. The long term persistence of the species in Southern England depends on the continued occurrence of conditions favouring the production of viable seeds and the recruitment of sufficient plants to offset yearly losses. This is not achieved every year. A series of consecutive unfavourable years would lead rapidly to the extinction of the population and improved habitat management may only serve to delay what environmental conditions make inevitable. The yearly variation in the ability of the population to maintain itself, may reflect the sensitivity of *O. sphegodes* populations to climatic factors in this part of its geographic range.

The sensitivity of the population to flowering plant mortality and reproductive performance, is due to the virtual absence of vegetative propagation in *O. sphegodes* (Hutchings 1987a). Management regimes which reduce flowering plant mortality are likely to be most successful in promoting the conservation of the species. Grazing immediately before and after flowering should be avoided. Increases in vegetative plant mortality appear to be relatively unimportant in affecting the number of plants in the population. However, this conclusion needs to be treated with some caution since the extent of plant mortality occurring between October and April is unknown (Hutchings 1987a). The sensitivity of population growth to dormant plant mortality illustrates the importance of the subterranean phase in the life cycle of orchids. The length of this phase, prior to the first emergence of plants above ground, is unknown in *O. sphegodes*. Mortality of individuals during this prolonged phase, in which the developing plant forms a symbiotic relationship with a mycorrhizal fungus, will limit the number of recruits (Wells 1981). The mortality of dormant plants may be influenced by factors such as waterlogging, physical disturbance, frost damage and predation. The frequency of factors such as physical disturbance and waterlogging could be reduced by careful habitat management, while the impact of low winter soil temperatures may be reduced if care is taken to ensure that site vegetation cover is not reduced excessively by autumn or winter grazing.

The model indicates that the effective reproductive performance of flowering plants severely limits population growth in this species. Summerhayes (1951) and Lang (1980) report that despite the ability of *O. sphegodes* to self pollinate, only 6–18% of seed capsules ripen in Great Britain, compared with 45% of plants on the Continent. This difference may be due to greater pollinator availability on the Continent. It would be an interesting experiment to see if localised artificial cross pollination affected the production of ripe seed capsules.

The value of the results of the model depends on the validity of the model developed. The model presented here is clearly too simple to provide a detailed and realistic picture of the complex population dynamics of *O. sphegodes*. Transition matrix elements are assumed to reflect biological constraints specific to the site and the species studied. It would be extremely interesting to know whether the transition probabilities of *O. sphegodes* plants on the Continent are similar to those obtained for the Castle Hill population. Although faults might be found in the way in which the subterranean phase of the life cycle and the reproductive performance of individual plants have been included in the model, this study does illustrate the value and use of simple models, which may be developed from a minimum of demographic data for conservation. The type of data required can be collected easily if a population is censused at regular intervals. From such data a transition matrix may be constructed and used to investigate the likely success or failure of different management approaches which alter the parameters of the model. If recruitment rates can be estimated then the model may also be used to assess the sensitivity of the population to mortality.

References

Begon, M. & Mortimer, M. (1986). *Population Ecology*. Oxford: Blackwell Scientific Publications.

Burden, R.L., Faires, J.T. & Reynolds, A. (1981). *Numerical Analysis*. Boston: Prindle, Weber & Schmidt.

Hutchings, M.J. (1983). Plant diversity in four chalk grassland sites with different aspects. *Vegetatio*, 53, 179–89.

Hutchings, M.J. (1987a). The population biology of early spider orchids, *Ophrys sphegodes* Mill. I. A demographic study from 1975 to 1984. *J. Ecol.*, 75, 711–27.

Hutchings, M.J. (1987b). The population biology of early spider orchids, *Ophrys sphegodes* Mill. II. Temporal patterns in behaviour. *J. Ecol.*, 75, 729–42.

Jeffers, J.N.R. (1978). *An Introduction to Systems Analysis*. London: Edward Arnold.

Lang, D. (1980). *Orchids of Britain*. Oxford: Oxford University Press.

Searle, S.R. (1966). *Matrix Algebra for Biological Sciences*. New York: Wiley.

Summerhayes, V.S. (1951). *Wild Orchids of Britain*. London: Collins.

Wells, T.C.E. (1981). Population ecology of terrestrial orchids. In *The Biological Aspects of Rare Plant Conservation*, ed. H. Synge, pp. 281–95. Chichester: Wiley.

Predicting the probability of the bee orchid (*Ophrys apifera*) flowering or remaining vegetative from the size and number of leaves

Introduction

There is a growing awareness among plant ecologists that the size of an individual is more important in determining its behaviour than its chronological age. Rabotnov (1950) was among the first to demonstrate that in any closed community there is likely to be a distribution of plants in different age classes. He noted that there would be seedlings, juveniles, immature adult plants, reproductive plants, vegetative adult plants and senescent non-flowering plants of great age, but he was unable to identify the factors which contributed to a plant switching from a vegetative to a reproductive state. More recently, Werner (1975), Baskin & Baskin (1979) and Gross (1981) have shown that for a number of biennials a minimum size must be reached before flowering can be induced and above a minimum size the probability of an individual flowering increases directly with rosette size.

This study focuses on the behaviour of rosettes of *Ophrys apifera* L. over a six year period, with particular reference to the fate of rosettes (flowering or remaining vegetative) relative to their age, size and number of leaves in any particular growing season.

Site details

The study area was a gentle, north-facing slope situated in Com's Field at National Grid Reference (NGR) 52/200795, about 600 m west of Monks Wood Experimental Station. Prior to 1960, this field had been part of a mixed farm. The present grassland was sown in May 1964 as a mixture of *Lolium perenne* L., *Dactylis glomerata* L. and *Trifolium repens* L., but now consists of a mixture of unsown grasses and forbs, with small amounts of the sown species. Since 1964, the grassland has been cut annually for hay, and except for 1970, when it received a top dressing of nitrochalk, no fertilisers or herbicides have been applied. A hedge was planted along the western edge of the study area in 1963 and is now about 6 m high.

The soil, derived from the Chalky Boulder Clay overlying Oxford Clay, is imperfectly drained and water lies on or near the surface for most of the winter.

During the summer, the soil dries and cracks, fissures 10 cm deep and up to 2 cm across being not uncommon. The pH of the top 10 cm was 7.8–7.9 (mean of 5 core samples taken in December 1985: determined on air-dried soil by glass electrode in a 1:2.5 soil:water paste).

Method of study

An area 10×10 m, known to contain a population of *Ophrys apifera*, was selected for study in July 1979. Hardwood pegs were driven into the ground at 5 m intervals along the western edge of the study area, about 1 m from the hedge. Each peg was labelled and cup-hooks screwed into the tops of the pegs. At each recording, the position of individual orchid plants in relation to the pegs was plotted, using surveyors tapes. Each orchid received a unique coordinate which enabled us to return to the same plant at successive recordings.

The whole population was recorded on 22 occasions during the period 1979–85. The main census was made in July each year when flowers were fully open. For each plant the following data were recorded:

1. state of plant, whether flowering or vegetative;
2. height of inflorescence;
3. the number of flowers on each inflorescence;
4. the state (alive, dead or damaged), and
5. the number of leaves.

Counts of the number of leaves in the basal rosette and measurements of the length and breadth of the longest leaf were made at intervals throughout each year, the actual timing of these observations depending on when leaves emerged above ground (September–December) and when weather conditions during the winter permitted. Plants were censused in September 1979, 1981; October 1980, 1982, 1983; November 1984; February 1980; March 1983, 1984; April 1981, 1982, 1985; May 1980, 1983, 1984 and every year in July, from 1979–85.

A population of *O. apifera* adjacent to the main study area was destructively sampled during one growing season (September 1983–June 1984) to provide more detailed information on the relationship between plant size and the probability of an individual flowering. In September 1983, 100 plants were marked with numbered labels as the leaves emerged above ground. On November 16 1983, March 7, May 17 and June 28 1984, 20 plants were selected on each occasion using a 12 cm diameter cylinder knocked into the ground to a depth of about 10 cm. Orchids within each core sample were carefully washed out, photographed and separated into leaves, shoots, roots, old tubers, new tubers and in those plants with inflorescences, into scapes and flowers. Leaf area was estimated by placing leaves between sheets of glass and tracing the outline onto paper, thereafter digitising the leaf shape and calculating the area. The volume of new and old tubers was estimated by water displacement in an Archimedes bottle. All plant parts were dried to constant weight at 85 °C in a forced-draught oven. Shoot apices were dissected to look for young inflorescences, and where these were present were accepted as an indicator that the plant would have flowered. The number of flowers on each inflorescence was counted and used as an estimate of potential flower production.

Results

General phenology

Leaves of *O. apifera* usually emerge above ground from early September onwards, but in some years e.g. 1985, appearance above ground may be delayed until November or even early December. Late emergence appears to be correlated with low rainfall in the period August to October.The greyish-green, elliptical-oblong shaped leaves which constitute the basal rosette increase in number and size gradually. Leaves remain green throughout the winter and spring and seem unaffected by low temperatures or snow cover, although fully expanded leaves may become blackened at the tips and edges by severe frosts in the absence of a snow cover. Severe blackening of individual leaves sometimes accompanied by the death of individual leaves occurs during periods of drying wind in late Spring, and appears most severe in small plants when all of the leaves may be killed, although the subterranean parts of the plant survive.

Tubers and roots are replaced annually. New tubers arise in November as a small white protruberance on the stem above the old tuber, eventually bursting through the leaf sheath, the yellow meristematic tip of the ellipsoidal-shaped tuber being especially prominent at this stage. Growth of the new tuber during the winter months is slow, but in mid-March the new tuber begins to increase in size rapidly and by late May the new egg-shaped tuber is as large as or larger than the old tuber and by the time the plant flowers in late June or July the old tuber has begun to shrivel as its food reserves are used up in flowering and new tuber formation.

White adventitious roots are visible as protruberances beneath the leaf-sheaths above the tuber in September. By mid-November these have burst through the leaf sheath and are up to 2 cm long. Large plants have 6–11 such roots by April; smaller plants have few roots, but tubers as small as 3 mm in diameter usually have at least one root. Roots become brown and necrotic in the month before flowering and are usually dead and shrivelled by late July. A generalised phenology, based on observations made on excavated plants is shown in Figure 1.

Relationship between leaf number, length of longest leaf and flowering performance

The number of leaves in the basal rosette increases steadily during the annual growth cycle, although there were large differences between individuals in the

Figure 1. Generalised phenology of *Ophrys apifera*, illustrating tuber formation and annual production of leaves, roots and inflorescence.

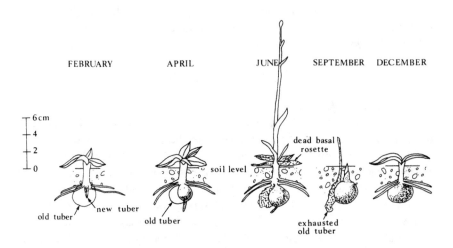

Figure 2. Frequency of leaf number, as a percentage of vegetative (open column) and flowering plants (hatched columns), in a population of *Ophrys apifera*, 1980–85. Counts were made when leaf numbers were at their highest.

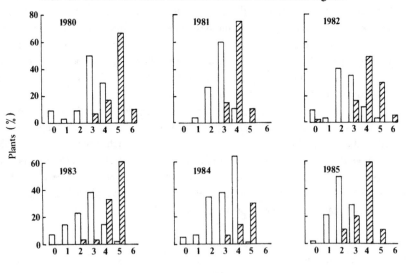

number of leaves in the rosette, probably reflecting differences in the age of the plants. During the six year study (1980–85), the mean number of leaves per plant for the whole population increased from 1.03 in late September to 1.98 by the end of November, reaching 3.23 in February, thereafter increasing to 4.05 in mid-April, reaching a maximum of 4.7 in mid-May. These average values mask considerable variation in the frequency distribution of leaf number between years (Figure 2), and between plants which flower and those which remain vegetative. In general, plants with most leaves and the longest leaves, and hence the largest leaf area, were more likely to flower than plants with fewer and smaller leaves.

The probability of *Ophrys apifera* flowering or remaining vegetative, as a function of leaf number in April or May, is given in Table 1. Plants with six leaves are certain to flower, plants with 0 or 1 leaf are certain not to flower (the single plant in April 1982 which apparently flowered without a leaf, probably had leaves which were grazed by molluscs and was erroneously recorded as leafless). In 1980, 1981 and 1985 all plants with five leaves flowered, and in the three other years plants with five leaves had probabilities of flowering of 0.93, 0.95 and 0.83 respectively. In four years out of six all plants with two leaves remained vegetative, the probability of plants with two leaves flowering in 1983 being 0.29 and 0.03 in 1985. The mean probability of a plant with three leaves flowering is 0.11, whereas a plant with four leaves has a probability of flowering of 0.66.

Plants which were going to flower in June had significantly more leaves in the basal rosette (P<0.001) than plants which were to remain vegetative in any month preceding June in which comparisons were made (Table 2). Differences were found as early as September 25 in the preceding year (1.49 leaves in plants which were to flower compared with 0.58 leaves in plants which did not flower; significant at P<0.001). Similar significant differences between plants which were to flower and those which remained vegetative were found when the length of the longest leaf was measured (Table 2), in all years and months, except for April 1981.

Changes in leaf area, leaf number, number of roots, tuber volume and weight, and total plant dry weight in the population destructively sampled from November 16 1983 to June 28 1984

There were highly significant differences (P<0.001) in all attributes measured at all sampling dates (except for new tuber weight in June) between plants which were going to flower and those which were to remain vegetative (Table 3). Plants which were going to flower in June had more leaves, a larger leaf area, more roots, larger tubers and a greater total plant dry weight in November, and at the three other sampling times (March, May and June), than plants which were to remain vegetative (non-flowering).

There was a progressive increase in all measured attributes with time, some growth taking place even during the winter months when mean daily temperatures were

Table 1. *The probability of Ophrys apifera flowering in July (F) or remaining vegetative (V) as a function of leaf number, measured in April or May, in the period 1980–85.*

No. of leaves	May 1980		April 1981		April 1982		May 1983		May 1984		April 1985		Total number of observations
	F	V	F	V	F	V	F	V	F	V	F	V	
0	0	1.00	–	–	0.25	0.75	0	1.00	0	1.00	0	1.00	15
1	0	1.00	0	1.00	0	1.00	0	1.00	0	1.00	0	1.0	30
2	0	1.00	0	1.00	0	1.00	0.29	0.71	0	1.00	0.03	0.97	102
3	0.10	0.90	0.08	0.92	0.37	0.63	0.05	0.95	0.04	0.96	0.10	0.90	143
4	0.33	0.66	0.71	0.29	0.84	0.16	0.57	0.43	0.55	0.45	1.00	0	106
5	1.00	0	1.00	0	0.93	0.07	0.95	0.05	0.83	0.17	1.00	0	64
6	1.00	0	–	–	1.00	0	–	–	–	–	–	–	5

Table 2. *Mean number of leaves and mean length of longest leaf in rosettes of Ophrys apifera among individuals which flowered or remained vegetative, on 14 occasions, 1980–85.*

Date	n		Mean number of leaves ± S.E.			Mean length of longest leaf (cm) ± S.E.		
	Flower	Vegetative	Flower	Vegetative	P<	Flower	Vegetative	P<
Feb 26 1980	94	83	3.23 ± 0.06	2.33 ± 0.07	0.001	ND[a]	ND	
May 14 1980	95	83	4.52 ± 0.08	3.37 ± 0.07	0.001	6.43 ± 0.15	5.93 ± 0.17	0.05
Nov 28 1980	59	120	1.98 ± 0.10	1.29 ± 0.08	0.001	5.03 ± 0.27	3.28 ± 0.22	0.001
April 7 1981	78	135	4.05 ± 0.08	3.10 ± 0.05	0.001	5.55 ± 0.16	5.23 ± 0.10	NS[b]
Sept 25 1981	146	74	1.49 ± 0.06	0.58 ± 0.07	0.001	2.74 ± 0.16	0.98 ± 0.15	0.001
April 19 1982	163	79	4.38 ± 0.06	2.91 ± 0.10	0.001	ND	ND	
Oct 26 1982	119	138	1.87 ± 0.05	1.05 ± 0.05	0.001	4.42 ± 0.17	2.86 ± 0.15	0.001
March 8 1983	124	148	2.99 ± 0.05	1.97 ± 0.06	0.001	5.32 ± 0.11	3.94 ± 0.12	0.001
May 23 1983	125	153	4.70 ± 0.08	2.84 ± 0.09	0.001	ND	ND	
Oct 1983	74	192	2.08 ± 0.03	1.38 ± 0.05	0.001	4.55 ± 0.17	3.19 ± 0.30	0.001
March 6 1984	74	196	3.04 ± 0.06	2.04 ± 0.06	0.001	5.33 ± 0.15	4.04 ± 0.24	0.001
May 1984	75	205	4.27 ± 0.04	2.81 ± 0.09	0.001	8.67 ± 0.31	6.29 ± 0.55	0.001
Nov 1984	46	218	2.02 ± 0.01	1.25 ± 0.03	0.001	5.16 ± 0.10	3.22 ± 0.17	0.001
April 1985	47	244	3.55 ± 0.09	2.23 ± 0.04	0.001	ND	ND	

[a]ND, not determined
[b]NS, not significant.

Table 3. Leaf area, number of leaves and roots, volume of old tubers, weight of new tubers and total plant dry weight of Ophrys apifera at four sampling dates, separated into plants with inflorescences (F) and plants without inflorescences (V). ($\bar{x}$ = mean; S.D. = standard deviation; n = number of plants in sample)

		November 16 1983			March 7 1984			May 17 1984			June 28 1984		
		V	F	P	V	F	P	V	F	P	V	F	P
Leaf area (cm²)	$\bar{x}$	1.92	17.69	<0.001	3.85	22.97	<0.001	6.62	28.64	<0.001	10.40	40.97	<0.001
	S.D.	1.49	8.34		2.67	8.00		7.98	14.37		7.70	13.10	
	n	7	23		9	20		11	23		15	13	
number of leaves	$\bar{x}$	1.00	2.45	<0.001	1.94	3.80	<0.001	2.45	4.67	<0.001	2.55	5.67	<0.001
	S.D.	0	0.60		0.58	0.59		0.82	1.12		0.86	0.75	
	n	7	23		9	20		11	23		16	13	
number of roots	$\bar{x}$	2.14	6.58	<0.001	2.77	6.80	<0.001	3.00	7.08	<0.001	4.31	7.23	<0.001
	S.D.	0.95	1.44		1.48	1.23		1.55	1.75		1.74	1.30	
	n	7	23		9	20		11	23		16	13	
Volume of old tuber (cm³)	$\bar{x}$	0.014	0.182	<0.01	0.015	0.746	<0.001	0.136	0.538	<0.001	0.141	0.723	<0.001
	S.D.	0.009	0.184		0.023	0.315		0.101	0.279		0.135	0.292	
	n	7	23		9	20		10	22		14	12	
Dry weight new tuber (mg)	$\bar{x}$	–	0.739	–	–	0.450	–	13.5	101.4	<0.001	101	108	NS
	S.D.	–	1.572		–	0.604		9.0	68.2		74	52	
	n	–	25		–	20		8	25		16	13	
Total plant dry weight (mg)	$\bar{x}$	12.16	174.30	<0.001	36.66	228.00	<0.001	76.75	313.69	<0.001	160.0	642.0	<0.001
	S.D.	6.33	100.43		17.51	85.44		66.14	169.52		122.0	251.0	
	n	6	23		6	20		8	25		16	13	

considerably below the 6 °C threshold which is commonly taken to be the temperature below which growth ceases. The mean daily temperature for the period November 16 1983 to March 7 1984 was 4.18 °C, yet in that period mean total dry weight of plants which were to flower increased from 174.3 mg in November to 228.0 mg in March and in non-flowering plants from 12.2 mg per plant to 36.7 mg, indicating that some growth and accumulation of photosynthates had occurred. However, the largest increase in dry weight occurred in the period May 17 to June 28, plants which were to flower increasing from 313.7 mg per plant to 642.0 mg; non-flowering plants from 76.7 to 160.0 mg per plant. Mean daily temperature for that period was 12.8 °C.

Although new tubers were recognisable in mid-November on plants which were to flower, they were very small (mean dry weight 0.74 mg) and they remained small until mid-March (mean dry weight 4.5 mg), increasing greatly in size between March and May (mean dry weight on May 17, 101.4 mg), with only a small increase in weight thereafter (108 mg on June 28). Formation of new tubers on the smaller, non-flowering plants occurred much later than on plants which were to flower, small protruberances being visible in March. By May 17 they were still very small (13.5 mg dry weight per plant) but by June 28 they were of almost the same dry weight as those on plants which flowered (101 mg per plant cf. 108 mg). Soluble carbohydrates were high in March and May (73% and 70% respectively), falling to 20% by June 28, when plants were in full bloom.

Changes in flower number per inflorescence with time

Inflorescences, less than 5 mm long, were detected in the November sample of plants, the number of flowers on each inflorescence ranging from 5 to 10, with a mean of 7.6 S.D. 1.7. Flowers were clearly differentiated into recognisable floral parts even at this early stage, although all parts were colourless and without any of the pigments characteristic of the mature flower. In the 20 plants sampled 3.5 months later, the number of flowers did not differ significantly from the earlier sample (range 6–9, mean 7.9 S.D. 1.09). By May 17, the mean number of flowers per inflorescence had fallen significantly ($P<0.01$) to 4.3 S.D. 1.06, range 4–6, and by the time flowers on the whole inflorescence had opened (June 28) the mean number of flowers per inflorescence was 3.0 S.D. 1.1, range 3–5, a significant decline ($P<0.01$) from the previous sample. During the six year study, the mean numbers of flowers per inflorescence ranged in the permanent plot from 2.6 in 1984 to 3.7 in 1982 when most plants flowered. The small sample counted in the plot which was destructively sampled falls in the centre of this range and cannot therefore be considered in anyway "abnormal".

Discussion

Unlike most flowering plants, terrestrial orchids pass a considerable part of their juvenile life as an underground protocorm or mycorrhizome depending on the fungal symbiont for their nutrition. The time spent underground varies greatly from species to species but is always greater than a year and may be as long as 15 years (Summerhayes 1951), although doubts about the accuracy of such statements are increasingly being made (Moller 1987a,b; Wells & Kretz 1987).

Ophrys apifera is said by Summerhayes to produce its first leaf two years after germination, but the evidence for this statement is not known. In our excavations, we found plants with narrow linear leaves and with tubers about 2 to 3 mm long which appeared similar in size and morphology to those about eight months old grown in aseptic culture in our laboratory, but we have no direct way of determining the age of seedlings found in the field. In the permanent study area, we have rarely found small plants, probably because of the difficulty of distinguishing them from the leaves of grasses. Plants sampled in our study were therefore those with obovate to ovate leaves, usually with more than one leaf in the basal rosette, and probably considerably older than those with single, linear leaves.

Bearing this difficulty in mind, total rosette leaf area, leaf number or the length of the longest leaf all give a reasonable quantitative prediction of the probability of *O. apifera* flowering or remaining vegetative (Table 1 and Figure 2). Differences in these leaf attributes between plants which are going to flower and those which remain non-flowering are apparent soon after leaves are produced in autumn, enabling predictions to be made concerning reproductive performance at all stages in the annual cycle. *Ophrys apifera* rosettes have a minimum size requirement for flowering (at least 2 leaves in mid-May) but once this size is surpassed, the probability of flowering increases with increasing rosette size and number of leaves. Other studies, mostly on 'biennial' plant species show a similar gradual, as opposed to an abrupt, increase in the probability of flowering with increasing rosette size (Werner 1975; van der Meijden & van der Waals-Kooi 1979; Gross 1981; Gross & Werner 1983).

Seasonal emergence time may be an important factor in determining the size an individual reaches at the end of a growing season. Plants which emerged first tended to be those which eventually had the highest number of leaves, whereas plants which emerged later had smaller and fewer leaves.

It is clear from the destructive sampling that plants with most leaves and the largest leaf area also had large tubers and high numbers of roots. This suggests that the critical size requirement for flowering may reflect that a certain level of carbohydrate reserves has to be reached before the plant is able to respond to the flowering stimulus, but whether this is a photoperiodic or vernalization requirement is at present unknown.

The dissection of shoot apices revealed that inflorescences with flowers differentiated into recognisable floral parts were present on November 16 1983 when the first samples were taken. It may reasonably be inferred that flower primordia were initiated therefore some time previous to this, possibly weeks, or even months earlier, when the plants were leafless, or with only one leaf at the most.

Current photoperiod research suggests that the cue or stimulus for flowering is received through the green leaf or shoot and is transmitted chemically to the shoot tip, thereby initiating flower primordia (Vince-Prue *et al.* 1984). This raises some interesting points. If, as our observations show, the plant was below ground when flower primordia were initiated, it seems probable that the stimulus was received by the old plant with leaves and the stimulus transmitted to the new tuber currently formed in that year. That plant would flower the following year if resources were not limiting. In other words, inflorescences are initiated a year before the plant actually flowers, and whether the plant produces an inflorescence is dependent on the food manufactured via the leaves in the current year.

A second point of interest is that the full reproductive potential of the plants was never achieved during the six year study at the Monks Wood site. Dissection of apices (Table 4) clearly showed that inflorescences with as many as nine flowers were present in November and March (mean 7.61 and 7.85 respectively) but that when the population flowered in June, the mean number of flowers per spike was 3.07 S.D. 1.11, with a maximum of five flowers per spike. In the six year study, in which 684 inflorescences have been measured and flowers counted, the mode for five years was three flowers per spike, and for one year (1982), four flowers per spike. In a four year study of a population of *O. apifera* growing on a roadside verge at Thurleigh, Bedfordshire (Wells & Farrell unpublished) the mode for number of flowers per spike was three, with a single plant having six flowers (n=56), behaviour similar to that shown by the Monks Wood population. However, plants with 10 and 11 flowers per spike have been recorded from a few sites; for example, in a population growing in a dune slack at Berrow, Somerset (pers. comm. Miss C. Saunders). We suggest that the plants at Monks Wood and Thurleigh are growing in sub-optimal conditions caused by insufficient nutrients or available water, with the result that flowers abort before valuable resources are channelled into reproductive structures which will not be able to produce seed.

Although the regulation of maternal investment in flower and seed production by aborting juvenile fruits or by altering the number of flowers produced has received scant attention from ecologists, our results and interpretation accord with those of Stephenson (1984) who showed that in the legume *Lotus corniculatus*, a restriction on resources limited its reproductive output by altering the numbers of flowers and inflorescences produced. The stage in the development of the inflorescence of *O. apifera* at which flowers are aborted probably depends on prevailing environmental conditions and almost certainly varies from year to year. Under extreme conditions, especially of drought, whole populations are known to have not produced a single inflorescence (e.g. at Milton, Cambridgeshire in 1980, pers. comm. R. Day), the leaves dying before the inflorescence had emerged from the basal rosette. The plants, however, survived as dormant tubers. At other times, usually following a dry period,

Table 4. *The number of flowers or flower primordia detected on developing inflorescences in a population of* Ophrys apifera *sampled on November 16 1983, May 17 and June 28 1984. Plants were selected at random at each sampling date from a population of 100 plants marked in November 1983. Figures in brackets denote uncertainty as to whether the structure observed under the microscope was a flower.*

Nov 16 1983	March 7 1984	May 17 1984	June 28 1984
7 (2)	8	3 (1)	2
5 (1)	9	3	4
5(1)	9	5	2
5 (2)	9	2	4
7 (3)	6	3 (1)	2
5	7	4	3
5 (1)	9	5	2
5 (2)	6	4 (2 aborted)	2
6 (2)	9	4 (3 aborted)	5
9 (1)	9	4	5
7 (2)	7	4	3
6 (2)	6	3 (1 aborted)	3
6 (2)	8	4	3
4 (1)	9	4	—
6 (1)	8	4	—
3 (2)	8	4	—
6 (1)	7	3 (2)	—
8 (2)	7	3 (1)	—
7 (2)	8	2 (2)	—
8 (2)	8	3	—
6 (2)	—	3 (1)	—
7 (2)	—	4 (1)	—
5 (1)	—	3 (2)	—
Mean 7.61	7.85	4.30	3.07
S.D. 1.69	1.09	1.06	1.11

inflorescences have been known to shrivel and die a few weeks before the flowers were due to open.

We conclude that leaf area and numbers of leaves are reliable predictors of the potential of a plant of *O. apifera* to flower or to remain vegetative. However, other factors, of which water supply at certain critical times is probably of most importance, may prevent the plant from flowering, or restrict its performance by causing the abortion of some, or all, of the flowers.

References

Baskin, J. & Baskin, C.C. (1979). Studies on the autecology and population biology of the weedy monocarpic perennial, *Pastinaca sativa. J. Ecol.*, **67**, 601–10.

Gross, K.L. (1981). Predictions of fate from rosette size in four "biennial" plant species: *Verbascum thapsus, Oenothera biennis, Daucus carota,* and *Tragopogon dubius. Oecologia* (Berl.), **48**, 209–13.

Gross, R.S. & Werner, P.A. (1983). Probabilities of survival and reproduction relative to rosette size in the Common Burdock (*Arctium minus*: Compositae). *Am. Midl. Nat.*, **109**, 184–93.

Meijden, E. van der & Waals-Kooi, R.E. van der (1979). The population ecology of *Senecia jacobaea* in a sand dune system. I. Reproductive strategy and the biennial habit. *J. Ecol.*, **67**, 131–53.

Möller, O. (1987a). Die subterrane Innovation und Wachstumszyklus einiger Erdorchideen. *Die Orchidee*, **38**, 13–22.

Möller, O. (1987b). Zur Notwendigkeit einer Renaissance der Erdorchideenkunde? *Die Orchidee*, **38**, 71–6.

Rabotnov, T.A. (1950). Life cycles of perennial herbage plants in meadow communities. *Proc. Komarov Bot. Inst. Akad. Sci. USSR.*, Ser. 3(6), 7–240 (in Russian).

Stephenson, A.G. (1984). The regulation of maternal investment in an indeterminate flowering plant (*Lotus corniculatus*). *Ecology*, **65**, 113–21.

Summerhayes, V.S. (1951). *Wild orchids of Britain.* London: Collins.

Vince–Prue, D., Thomas, B. & Cockshull, K.E. (1984). *Light and the Flowering Process.* London: Academic Press.

Wells, T.C.E. & Kretz, R. (1987). Asymbiotische Anzucht von *Spiranthes spiralis* (L.) Cheval. vom Samen bis zur Blüte in fünf Jahren. *Die Orchidee,* **38**, 245–7.

Werner, P.A. (1975). Predictions of fate from rosette size in Teasel (*Dipsacus fullonum* L.). *Oecologia* (Berl.), **20**, 197–201.

British orchids in their European context

Of the estimated 191 species (Baumann & Künkele 1982) currently accepted for Europe, The Middle East and North Africa, only 53 are native to the British Isles. One of these, *Spiranthes aestivalis* (Poir.) L.C. Rich., is generally considered to be extinct.

Many Continental species, e.g. *Epipactis muelleri* Godfery, *Limodorum abortivum* (L.) Sw., *Orchis coriophora* L. and *Serapias cordigera* L., were prevented from ever reaching our shores by the formation of the English Channel around seven thousand years ago. Others, such as most *Ophrys* species, e.g. *O. scolopax* Cav. and *Serapias* species, e.g. *S. lingua* L., require a warmer, drier climate and only extend as far as south-west and south-central France. Although the majority of British orchids may be seen in greater numbers across the Channel, it is a mistake to assume that our native orchid flora is merely a poor representation of Continental Europe. Several species, particularly those of chalk grassland, are as widespread in Britain as elsewhere. Among these may be included *Anacamptis pyramidalis* (L.) L.C. Rich., *Dactylorhiza fuchsii* (Druce) Soó and *D. maculata* (L.) Soó, *Epipactis helleborine* (L.) Crantz, *Gymnadenia conopsea* (L.) R.Br., *Orchis mascula* L. and *Platanthera chlorantha* (Custer) Reichb.

A few species have their main European distribution in the British Isles, e.g. *Epipactis phyllanthes* G.E. Sm., a variable plant, some forms of which have cleistogamous flowers. On the Continent it is only found in western France and Denmark. *Spiranthes romanzoffiana* Cham. is an example of a North American species which, in Europe, is confined to Ireland, Western Scotland and Devon. *Epipactis dunensis* (T. & T.A. Stephenson) Godfery, from the dune systems of north west England and Anglesey, is one of the very few orchids probably endemic to the British Isles. Other interesting species include *Dactylorhiza praetermissa* (Druce) Soó, restricted to southern Britain, northern France, the Low Countries, Denmark and one isolated site in northern Italy, and *Epipactis leptochila* Godfery restricted to southern Britain and scattered localities in central Europe. Even more interesting is *Neotinea maculata* (Desf.) Stearn, a southern European species widespread in Mediterranean countries, but occurring in Eire and the Isle of Man. The nearest Continental localities are in Spain and Portugal and it belongs to the so-called 'Lusitanian element' of the British flora.

Other taxa are at the edge of their natural range in the British Isles, such as *Ophrys sphegodes* Miller and *O. holoserica* (Burm. f.) W. Greuter (syn. *O. fuciflora* (Crantz) Moench), *Himantoglossum hircinum* (L.) Sprengel and *Cephalanthera rubra* (L.) L.C. Rich. *Orchis laxiflora* Lam. extends to the Channel Islands but never crossed the English Channel to reach southern England.

Threatened British species

A large proportion (about 25%) of our present orchid flora is either rare or endangered. Many of these are the species at the edge of their distribution in the British Isles. It is interesting, therefore, to look through the Kew herbarium collections prior to 1970 and references in the literature to see how many species regarded as very rare today were formerly much more widespread. Many of these are common or widespread on the Continent, although even these are increasingly vulnerable in many areas. A few examples are given below.

Cypripedium calceolus *L., Lady's Slipper Orchid*

This was first recorded from Yorkshire in 1640. Sowerby, in his English Botany (1869), remarked that it was 'very rare and now nearly if not quite extinct ... occurred in several stations in Yorkshire as lately as 1849'. It was eradicated from its original site of discovery by a gardener in 1796 who found a ready sale for it. It had been recorded from many sites in six vice-counties in the north of England in the past, but today is found only in one closely guarded site. It is widespread in central and northern Europe, but declining in the more accessible locations. This is a vulnerable plant owing to its showy flowers and exotic appearance and is, perhaps, the one British species that conforms to the layman's traditional image of an orchid.

Ophrys holoserica *(Burm. f.) W. Greuter, Late Spider Orchid*

Once recorded from at least eleven localities in two vice-counties, today it is only found in Kent. It is widespread in western, central and southern Europe east to the Middle East.

Orchis militaris *L., Military or Soldier Orchid*

Once recorded from eleven localities in seven vice-counties, today present in two vice-counties only. It was recorded from Harefield in Middlesex during the last century. It is widespread throughout much of Europe east to the Caucasus.

Orchis simia *Lam., Monkey Orchid*

First recorded in 1660 and present in at least nine localities in four vice-counties, but today restricted to only two. It is widespread in western and southern Europe, North Africa, the Middle East and the Caucasus. Sites in central Europe are scattered.

Orchis purpurea *Huds., Lady Orchid or Maids of Kent*
First recorded from Kent in 1666 and once found in at least twenty-three localities in five vice-counties. Today it is probably confined to Kent. It is widespread throughout much of Europe, excluding parts of the north east, as well as Algeria, Turkey and the Caucasus. Flower colour and lip shape are variable.

Epipogium aphyllum *Sw., Ghost Orchid or Spurred Coral-root*
First recorded from Herefordshire in 1854. It was recorded from four vice-counties in the past, but today occurs erratically only in one or two sites in the Chiltern Hills. This is the most widespread of our examples, ranging from Europe across northern Asia to Japan. In Europe it is scattered, although often locally abundant, but never common over large areas.

Himantoglossum hircinum *(L.) Sprengel, Lizard Orchid*
First recorded from Kent in 1641. It was recorded from twelve vice-counties in the past and still flowers regularly in three and sporadically in four. It is distributed in western and central Europe and North Africa.

Dactylorhiza traunsteineri *(Sauter) Soó, Narrow-leaved Marsh Orchid*
Recorded from twenty vice-counties in the past and today found in eleven vice-counties, but, since plants are easily overlooked, it may well be found to be more widespread. It occurs in scattered sites in the alpine regions of central Europe and in Scandinavia.

Spiranthes aestivalis *(Poir.) L.C. Rich., Summer Lady's Tresses*
First recorded in 1840. The classic site was in the New Forest in Hampshire where, in 1900, nearly two hundred plants flowered. By the early 1930s only twenty remained. It was also recorded from the Channel Islands, but is now thought to be extinct both there and in Hampshire. It is distributed in central and southern Europe, North Africa, with isolated sites in the Low Countries and Yugoslavia.

Causes of increased rarity of species
What are the reasons for the rarity of these orchids in the British Isles? First of all, the rapid and widespread changes in agriculture and land use in the last one hundred and fifty years. Much of our species-rich chalk downland habitat was lost when it was ploughed up for agriculture during World War Two. The decline of the rabbit population caused by myxomatosis has led to a subsequent increase in scrub invasion on many remaining chalk downs. Chalk and limestone areas with their associated rendzina soils support much the largest number of orchid species. This is one of the reasons why France and Greece have such a diverse orchid flora. Here there are still large tracts of 'unimproved' calcareous habitat. Species such as *Orchis*

militaris L., *O. purpurea* Huds. and *O. simia* Lam. are still relatively common in France but European Economic Community (E.E.C.) agricultural policy has led to their rapid decline in some areas there also. Even roadside verges are rapidly disappearing in the intensively farmed north of France.

The 'improvement' of old species-rich meadows and the widespread application of weed killers and artificial fertilisers have all contributed to the decline of once familiar species. Certain orchids are naturally scarce owing to their specialised habitat requirements and fungal associations, e.g. many wetland species and saprophytes. Mycorrhizal associations are thought to be adversely affected by treatment with artificial fertilisers. Species such as *Orchis morio* L. often grew in hundreds in old pastures mixed with cowslips and other formerly commonplace flowers. Many orchids have also declined in numbers due to drainage of fens, marshes and traditionally managed water meadows. The extinct *Spiranthes aestivalis* (Poir.) L.C. Rich. and the very rare *Liparis loeselii* (L.) L.C. Rich. are good examples of wetland species now declining throughout Europe. The British also have a mania for 'tidying up' the countryside, and unnecessary hedgebank and verge trimming in particular seems to be a too often practised vice. The second reason for rarity in the British Isles is three hundred years of overcollecting by gardeners and botanists alike.

The need for conservation

What is the point of orchid conservation, particularly when so many species are at the edge of their natural ranges in the British Isles anyway and others are relatively abundant elsewhere in Europe? It hardly seems worthwhile bothering. There are several good reasons.

Some orchids such as *Epipactis dunensis* (T. & T.A. Stephenson) Godfery and *Spiranthes romanzoffiana* Cham., mentioned earlier, which are only found in the British Isles, are obviously in need of special protection. British populations of species widespread in Europe, however, may also differ in slight details from their Continental neighbours. Populations, particularly of widely distributed species, often vary considerably over a given geographical area. Some may be distinct enough to warrant separate subspecific status, e.g. the endemic *Dactylorhiza majalis* (Reichb.) P. Hunt & Summerh. subsp. *scotica* E. Nelson. The British form of *Orchis simia* Lam. was said by John Lindley (1835) to differ slightly from Continental plants by its 'more slender habit, narrower few-flowered spikes and bluntish leaves' and was named *O. macra* Lindl. It is thought that the British lady's slipper orchid might have a different genotype from populations on the Continent. The introduction into the wild of foreign stock would alter this.

A final and perhaps more important reason for orchid conservation in the British Isles is that orchids can be used to heighten public awareness of habitat and plant conservation in general. They have a high profile and are, perhaps, the 'pandas of the plant world'. *Cypripedium calceolus* L. is politically worth its weight in gold for publicity purposes and to demonstrate what could happen to other rare plants. Less interest is likely to be generated over some obscure hawkweed or 'lesser spotted fleabane'!

Not all is gloom and doom however. In the British Isles we do have many fine colonies of *Dactylorhiza* which seems to thrive in our damp Atlantic climate. Man-made habitats such as disused chalk and gravel pits, quarries and motorway embankments are proving to be the species-rich habitats of today. Large colonies of *Dactylorhiza* often appear with surprising rapidity in suitable disturbed sites, e.g. power station slag heaps in Essex, where confusing hybrid swarms are not uncommon. Species such as *Dactylorhiza fuchsii* (Druce) Soó and *Listera ovata* (L.) R.Br. are now well established on motorway embankments. *Himantoglossum hircinum* (L.) Sprengel is actually increasing in numbers and *Cephalanthera rubra* (L.) L.C. Rich. has recently been discovered in Hampshire for the first time.

Much valuable work has been and continues to be done by local county naturalist trusts in maintaining reserves and it is hoped that further sites will be acquired in the future. Farmers and industrialists are today increasingly aware of the need for conserving habitats. The attitude of the general public can only be changed through subtle education and the opportunity of seeing wild orchids on reserves can only help.

At best conservation of habitats can only be a holding operation. Projects such as the pioneering micropropagation work in progress at Kew take things a stage further (Clements *et al.* 1986). If successful it will enable stocks of endangered species to be established in cultivation with the objective of future reintroduction into the wild. This, of course, will only be possible so long as suitable habitats remain.

As a postscript, a recent paper (Kenneth *et al.*, 1988) reports the addition of a new member to the British orchid flora, viz. *Dactylorhiza lapponica* (Laest. ex Hartman) Soó (syn. *D. pseudocordigera* (Neuman) Soó), a marsh orchid otherwise known from Scandinavia and the European Alps. The British populations, from northern and western Scotland, have been known for some time, but it is only recently that their true identity has been confirmed.

References

Baumann, H. & Künkele, S. (1982). *Die Wildwachsenden Orchideen Europas.* Stuttgart: Kosmos.

Clements, M.A., Muir, H. & Cribb, P.J. (1986). A preliminary report on the symbiotic germination of European terrestrial orchids. *Kew Bull.*, **41**, 437-45.

Kenneth, A.G. Lowe, M.R. & Tennant, D.J. (1988). *Dactylorhiza lapponica* (Laest. ex Hartman) Soó in Scotland. *Watsonia*, **17**, 37–41.

Lindley, J. (1835). *Synopsis of the British Flora.* London: Longman.

Sowerby, J. (1869). *English Botany*, vol. 9. London: Robert Hardwicke.

The Nature Conservancy Council and orchid conservation

Introduction

The Nature Conservancy Council (NCC) is the official government body concerned with the policies and practices of nature conservation. Since the passing of the Wildlife and Countryside Act in 1981 it has had a legal duty to fulfil its role in protecting sites and species of special interest in Britain.

Listed under Schedule 8 of the Act are 93 plants which are so rare that they are considered worthy of special protection. Nine of these are orchids (Table 1).

In Britain we have about 50 native, terrestrial orchid species. We say 'about' as there is constant debate regarding the exact status of several of our species. Twelve of these are nationally rare and are listed in the Red Data Book (Perring & Farrell 1983). Two other species, *Spiranthes aestivalis* and *Ophrys bertolonii*, are extinct and a third, *Hammarbya paludosa* is threatened in Europe, but thankfully more widespread in Britain.

In 1978, Lynne Farrell transferred from the Biological Records Centre, Institute of Terrestrial Ecology, at Monks Wood Experimental Station to take up a post entitled 'Botanist' in the Chief Scientist's Team of the NCC based at Huntingdon. This post had several facets, including grasslands and heathlands, as well as rare plants. Since then, the work has been directly concerned with rare plant conservation and particularly with orchid protection as they are a group of plants which have, and will continue to attract a great deal of attention. It is their floral structure, their delicate beauty, their mystique, their folk lore and their rarity which combine to make them the most intriguing and challenging genera we have. People travel miles to photograph and simply to see and enjoy these plants. Sometimes this becomes an obsession and 'orchid twitchers' are definitely on the increase. There is no wish to discourage the widespread interest in orchids, but practical protection measures must operate so that individual plants can continue to survive and multiply so that future generations can also have the pleasure of seeing these remarkable life-forms.

During 1982 it became obvious that several of our orchids were being vandalised. Plants of lizard, military, monkey, ghost, early spider, lady, early purple and bee orchid were dug up over a three year period. By whom remains a mystery. It may be that the people who did this were full of good intentions and wished to propagate them and increase their numbers – a highly commendable aim. However, as we are still

Table 1 *Orchid species listed under Schedule 8.*

Species name	Common Name
Cephalanthera rubra	Red helleborine
Cypripedium calceolus	Lady's slipper
Epipogium aphyllum	Ghost orchid
Himantoglossum hircinum	Lizard orchid
Liparis loeselii	Fen orchid
Ophrys holoserica	Late spider orchid
Ophrys sphegodes	Early spider orchid
Orchis militaris	Military orchid
Orchis simia	Monkey orchid

discovering, orchid germination and propagation is a complex process. The best chance of survival for these orchids, especially the rare ones, is to manage them properly in their native habitats. There are very good reasons why they are restricted to particular areas in the first place.

Orchid wardening scheme

On 6 October 1983 an *ad hoc* meeting was held at The Herbarium, Kew, of the people directly connected with British orchid conservation and representing various official bodies and societies. At this meeting all aspects were discussed, each person pooling their experience and knowledge. There was unanimous agreement on our future programme. Kew were to experiment with orchid germination and propagation, NCC were to identify suitable sites for re-introduction of species successfully grown, the Royal Society for Nature Conservation (RSNC) and NCC were to initiate an Orchid Wardening Scheme to protect the sites of the rarest species. This programme is a long-term strategy and is still evolving.

During the discussion, the relative merits of restocking, re-introduction and introduction were analysed. Whilst all had some purpose, it was generally agreed that re-introduction presented the best line of attack. This was because the appropriate strains of mycorrhiza necessary for successful growth were likely to be found at the sites where the orchid previously grew.

Following this, Dr Perring (RSNC) contacted the County Trusts for Nature Conservation who had rare orchids that had been damaged and Lynne Farrell approached NCC Regional Staff for their advice. The outcome was that the RSNC applied to NCC and World Wildlife Fund for a grant to employ orchid wardens in 1983. Over the summer, a total of 7 sites were wardened. The orchid species receiving this special attention were *Himantoglossum hircinum*, *Ophrys sphegodes*, *Orchis militaris* and *Orchis simia*.

The scheme was far more successful than we anticipated. No orchids were dug up or damaged at the wardened sites. Over 1000 people visited them and were told first-hand about the research that was being carried out and the conservation problems. Many became Trust members. The presence of a warden on site for several weeks, especially during the flowering period, enabled detailed biological observations to be made on topics such as insect visitors and pollination rates. Information on other groups such as butterflies, birds and small mammals was collated. But one of the more important spin-offs was the positive publicity that resulted. Orchid conservation is an attractive subject and the press were not slow to latch on to this topic.

Having a warden on a vulnerable site enables us to open the area to the general public for several weeks, rather than just on the occasional open day. People are much more appreciative if they are allowed to see rare plants for themselves, instead of it being declared a 'no go' area. Obviously some restriction has to be placed on visitors – the trampling effect of hundreds of feet does compact the soil adjacent to the flowering plants and seriously affects the chances of new seedlings becoming established. So we have often selected the most photogenic plants and those easily accessible as 'show plants'. This allows everyone to get as close as they want to a fine specimen, whilst ensuring the survival of the majority of the population.

The scheme has continued since its inception, widening its geographical coverage, so that in 1987 10 sites were wardened protecting 6 of our most vulnerable species – *Cephalanthera rubra, Corallorhiza trifida, Himantoglossum hircinum, Orchis militaris, Orchis simia* and *Spiranthes romanzoffiana*. In 1987 the practical wardening aspects have been co-ordinated by Nick Stewart, Conservation Officer for the Conservation Association of Botanical Societies (CABS). Each warden is visited on duty by one 'expert' who has been involved in the work from its beginning. As well as offering practical advice, they make sure each warden has a copy of the booklet 'Wardening of Rare Plants' (Guidelines for Wardens). A much more comprehensive text based on an original outline drafted in 1983 was produced in 1988 by Judith Church, a former warden, with help from other participants. This is available from CABS (323 Norwood Rd, London SE24 9AQ), and we hope will be of use for many species, not just for orchids.

What of the future? The success of the scheme means that it has perpetuated itself. However, it would be unrealistic to assume that it will continue in its present form for ever. One hopes that there will eventually be no need for a guardian of the orchids, as people will be better educated and realise that destroying these beauties is destroying part of their natural heritage – a heritage that we all need to play a part in caring for.

Orchid gardens

Whilst studying in the Netherlands for the Biological Flora of *Orchis militaris*, Lynne Farrell visited the area known as the Gerendal, near Maastricht, at the recommendation of Terry Wells. Dr Jo Willems, a Dutch botanist, who has worked on chalk grassland and made a detailed study of *Orchis simia*, accompanied her. This is a natural valley which is now set aside as a reserve where traditional farming methods are still used, old breeds of animals are maintained and old varieties of fruit are perpetuated. In addition to this is an 'orchid garden', a small area of limestone grassland approximately 0.5 ha in extent, which already contained several native Dutch orchid species. Other species have been 'added in' – rescued from sites that were to be destroyed. The individual plants are named and many are near to the footpath that winds through the 'garden' enabling visitors to examine them in the closest detail. There is a warden who conducts groups around and informs them of the project and the ecology of the different species. Unfortunately, no literature was available on site, which was a missed opportunity to spread the message of orchid conservation. Tens of thousands of visitors are received each year. The Dutch obviously care for the remnants of their natural heritage that they have left.

`We think that people also care in Britain, and we see the setting up of several orchid gardens as a logical next step in the orchid conservation programme. What we need now is the acquisition of suitable areas, preferably where orchids already exist and preferably on the chalk in Southern England where most of our species are found. Then we need wardens, possibly during the summer period only, to help manage these areas and to explain to visitors the unique qualities of these fascinating plants. A small entry charge and the sale of orchid literature will help to finance the warden. A successful parallel can already be seen with the butterfly houses that are now springing up near major towns throughout Britain. Many of the butterflies on display here are the spectacular tropical forms. Our native British orchids may not be as gaudy as their tropical rivals, but they do have fascinating life-histories, about which we are still learning, and they do have a certain intricate and delicate beauty. They are well worth our efforts to conserve.

Orchid gardens would not only enable many more people to see the plants but have the additional function of decreasing pressure on native sites. At the moment, our beautiful lady's slipper is reduced to one clump in Yorkshire. It is a truly spectacular flower and one can understand why people want to see it so badly. Unfortunately, visitor pressure in the early 1980s led to erosion of the slope nearby and to compaction of the soil around the clump itself. This limited the chances of vegetative spread and seedling development. Requests were issued in several widely read botanical journals not to visit the site. We must thank everyone who heeded these requests.

During the last four years the grassland has recovered, the clump has grown in stature and six seedlings have been found. Kew have been able to raise protocorms from seed and we await the development of green plantlets. The future for this one clump must still be precarious, but we are beginning to glimpse light at the end of the tunnel. We look forward to the day when it is possible to say – go to site X in order to see *Cypripedium calceolus*. The pleasure gained from simply observing its beauty should be enough to override concern as to whether that site is in the wild or in a 'garden'.

Rare Plants Survey Scheme

A Rare Plants Survey Scheme currently run by NCC also contributes to the present and future conservation of British orchids. Detailed historical studies of rare species are made in certain regions and this information applied to identifying and recording surviving sites. During this process useful information on the ecology of some rare orchids can turn up and management is discussed with local botanists and within the NCC local offices. Recently, co-operation between The Rare Plant Survey Scheme and the Kew Micropropagation Unit has resulted in new examination of fungi compatible with lizard orchids growing in Britain. The surveys are useful in focusing regional interest and effort on rare plant conservation and in identifying areas where rare orchids can take their invaluable educational and publicity role without endangering fragile populations.

References

Perring, F.H. & Farrell, L. (1983). *British Red Data Book: 1. Vascular Plants.* Lincoln: Society for the Promotion of Nature Conservation.

13 RICHARD C. WARREN

A private conservation project in the coastal rainforest in Brazil: the first ten years

Introduction

The coastal rain forest of Brazil is a long, narrow strip of tropical forest which has been deeply eroded by small-scale and large-scale agriculture over many years. Around the large conurbations of Rio de Janeiro, Sao Paulo and Belo Horizonte the damage is especially severe, but pockets of virgin forest still remain which are of interest and in need of conservation. We have been working over the past ten years on a 1000 hectare estate near to Novo Friburgo in the upper Macae valley. The object of the part-time project has been to create a self-supporting reserve with little or no destruction of natural resources except the use of specific woods for building purposes. To achieve this, three main fund raising methods have been used:

1. A farm has been constructed for the laying, hatching and rearing of pheasants, duck, partridge and guinea fowl.
2. Tour groups from the UK and USA have been invited to stay and study orchids in their natural habitat with ourselves as guides.
3. A business has developed in the UK to germinate, grow and sell orchid seed and seedlings, and to provide information collected in the field about lesser known orchid species.

Game Farm

After numerous problems, the farm now produces around 1000 guinea fowl and 1000 pheasants per annum. The problems arose from the lack of electricity for hatching large quantities of eggs and the unreliability of the substitute, gas. Commercial feed also proved to be irregular in composition and invariably treated with antibiotics, which we considered undesirable. Consequently, an area of land has now been given over to growing sufficient maize, yams and cabbage to make our own feed mix.

Tour groups

Groups of orchid growers and naturalists from the USA and the UK have visited the reserve for periods ranging from four days to three weeks, often returning.

All have expressed their satisfaction at what they were able to learn about orchid growing.

Orchid seedling business

This was set up in 1979 under the name Equatorial Plant Company. Although it was initially confined to the collection of seed from and information about plants of the Serra do Mar, it has since broadened to include a wide range of species of the genera *Cattleya* and *Laelia* and, and in collaboration with Mr Woods of the Royal Botanic Garden, Edinburgh, a range of high altitude species from New Guinea are also grown.

The objectives of the Equatorial Plant Co. are to promote lesser known orchid species to the amateur growers; to finance trips to Brazil to work in the conservation area and make collections; to study the possibilities of re-introducing species into the wild; and to promote the exchange of reliably documented seed-grown material with other laboratories.

Collection techniques

Seed has been collected either dry or in green pod, transported to the laboratory in the UK, and germinated using standard techniques. Where plants are particularly rare in the field, pollinia have been transported between specimens and hand pollinations made. Seed is infrequently found on many *Oncidium* spp. which are generally self-sterile. Hand pollination has resulted in the production of pods and seed.

After the collection of young pods too immature to transport to the UK, embryo culture has also been used. A transportable polythene cabinet allows aseptic work. Embryos from young pods of *Zygopetalum maxillare* were plated directly onto agar medium containing coconut milk, and the embryos were greening-up on return to the UK three weeks later. *Z. maxillare* was chosen because of its specific habit of growing on tree-fern and the difficulty of establishing mature plants in cultivation. Seed has also been collected from other species, returned to the UK and deposited at Botanic Gardens where seed has been raised at later flowering periods. The main objective of the collection and sampling methods has been to preserve the parent plants unless they are for botanical study. Non-destructive sampling is quite feasible and saves considerable time since embryos or growing seedlings collected in the field using aspetic techniques can be easily transported.

Species chosen for collection fall into three categories. Firstly, available specimens: if seed is abundant in nature then it is collected and stored for possible interest. Secondly, species that are rare in cultivation: many species from the Serra do Mar were well known in Victorian times and were represented in most collections. Nowadays with much interest in hybrid orchid material, such species as *Oncidium marshallianum*, *O. gardneri* and *O. enderianum* are infrequently seen in catalogues of commercial growers, so seed of these species has been set and germinated. Thirdly, seed raised from good specimens of popular orchids: *Sophronitis coccinea* var. *grandiflora* is an example of such a species. Frequently the commerical specimens are small flowered whereas the specimens from this area of the Serra do Mar have large well-formed flowers. Seed collected from good specimens in the wild have bred true, and introduced superior plants to commercial growing.

Newsletter

Since the species grown are frequently not well known commercially, growers, both private and commercial, need to be provided with information about their growth requirements, their habitat and climate in order to cultivate the plants with confidence. Information has been collected in the field in Brazil and concentrates on detail that will assist in culture of the plants concerned, or be of general scientific interest.

Figure 1 illustrates an observation made on the colonies of *Maxillaria acicularis*, a small epiphytic orchid of interest for terrarium culture. Superficially the colony is rootless except at the point of attachment to the substratum. Pabst & Dungs (1977) writing about the similar *Maxillaria madida* alliance mention that the pseudobulbs are rootless. If the colony is dissected around the distal part of the rhizome and the bracts removed, roots are found along the length of the rhizome to the most distal pseudobulb. Although these are not normally visible, they presumably function in the water relations of the plant. Figure 2 demonstrates some work on the excavation of part of a colony of *Zygopetalum mackayi* to examine the extent of the rooting system. The length and branching of roots over four separate years was measured and is described. A four-year-old rhizome has just under 50 metres of root which is branching, active and covers a huge area. Since a normal colony of *Z. mackayi* may have 6-10 rhizomes, the total root length of a colony up to four years old may extend to 500 metres. This information is of interest to a grower since the impression formed of pot-grown plants is that only the current year's roots are active, and older roots should be periodically cut off. The lesson learned from these observations is that species such as these which can have such a length of active root, may be better grown in beds rather than pots. Confining such plants in pots may explain why they rarely produce specimens as large or spectacular as are found in the wild.

Climatic observations are also appreciated by growers. Temperature maxima and minima of the plants, and details of natural fluctuations in humidity, are of interest. The author was once lost in the jungle at night and, forced to sleep in the forest, noticed that the canopy was soon sodden after dark and dripped all night through. This dripping during a quite cloud-free night followed a hot summer day. The impliction is that is may be better to keep a greenhouse wet at night, whereas the normal recommendation has always been to damp down during the day and not moisten after noon, or at least to ensure that all water has dried off before nightfall (Warren 1984).

Re-introduction of species in the wild

This work has taken two forms: the re-introduction to the area of species lost by previous overcollection in the last century; and the introduction of laboratory-grown seedlings. Many seedlings taken to Brazil in flasks were lost because the weaning process is so critical to later success, and supervision of the plants is not possible in the short amount of time allowed for each visit. However, laboratory-reared and nursery-established seedlings have a high measure of success, and one of the species which we have now re-established in flowering-sized colonies is *Laelia cinnabarina*. *Laelia pumila* seedlings have also been established. *Laelia crispa* colonies

Figure 1. Line drawing of *Maxillaria acicularis* with floral morphology (A) and a dissection of the rhizome to expose the roots (B). (Warren 1985a)

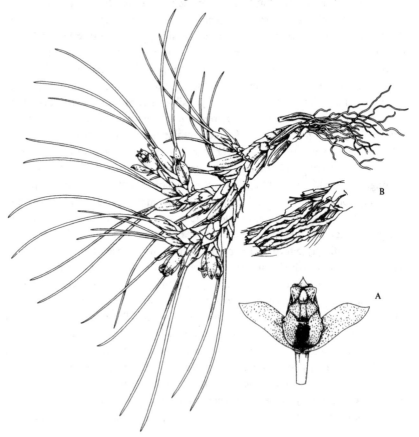

have been established as adult plants after collection from a neighbouring area of de-afforestation. A lorry load of plants was cut from the felled trees and replaced on a variety of 'host' trees. Best growth was seen on rough barked trees and on dead trees where the roots burrowed under the bark and extensively branched. During the first six months after transplantation, about 40 metres of roots were produced per colony, and most plants flowered within a year. Seed has been set by hand and allowed to dehisce naturally. After two years the first seedlings were found growing around the parent colony. Seedling growth is extremely slow; the seedlings only reached 5 cm after three years, which is equivalent to six months growth *in vitro* . It has not yet been observed whether natural pollination of the transplanted colonies has occurred since controlled seed production has been pursued as a priority.

Figure 2. Diagram of the root system of excavated colony of *Zygopetalum mackayi*. (Warren 1985c).

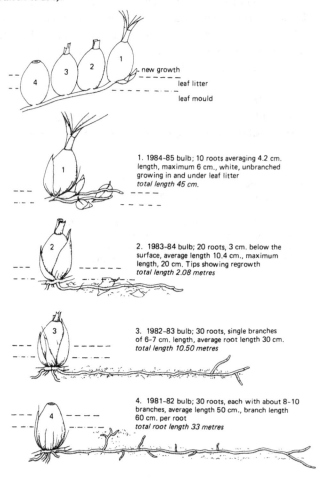

Future prospects

We have explored our land in the upper Macae quite thoroughly and noted about 250 species of orchid. We have now compiled extensive habitat notes, photographs and descriptions of many of these. We have now moved to a new area about 12 km away and 500 m lower in altitude. As is often the case in Brazil, ownership of land is wrapped in confusion. In this case the previous owner had disappeared, and the land was inhabited by squatters. However, the squatters were bought out (Warren 1985b) and a company, Sophronitis Enterprises Ltd, set up in a way that may not easily be dissolved. This new area is larger, has the source of the river Rio dos Flores, and is bounded on three sides by high, ridged hills. On the south side there is a government owned National Park, and the other neighbours are committed to conservation rather than exploitation. First examination suggests that the flora is richer than in our original area, since this south facing valley is warmer. Work to document as much of the plant life as possible is now in progress in collaboration with botanists at the Rio de Janeiro Botanic Gardens. The area is of sufficient size to support large mammals, and wild pig, puma and monkeys have been found. Since hunting with dogs has previously driven out much of the larger animal life, it is hoped that we will see a flourishing of the fauna.

Acknowledgements

I would like to thank David Miller whose enthusiasm and generosity has made this work possible; thanks also to Paddy Woods whose perseverance has been an inspiration.

References

Pabst, G.F.J. & Dungs, F. (1977). In: *Orchidaceae Brasiliensis*, Vol. 2, p 186. Hildersheim: Brucke Verlag.

Warren, R.C. (1984). *Newsletter*, Equatorial Plant Co., **1** (No.1), 3.

Warren, R.C. (1985a) *Newsletter*, Equatorial Plant Co., **2**(No. 2), 5.

Warren, R.C., (1985b) *Newsletter*, Equatorial Plant Co., **2**(No. 3), 9.

Warren, R.C. (1985c) *Newsletter*, Equatorial Plant Co., **2** (No. 4), 14

The role of the living orchid collection at Kew in conservation

Introduction

The living orchid collection at Kew consists of nearly 10 000 accessions (many of them represented by more than one plant) with around 370 genera and 3500 species represented. The collection comes under the Tropical department and is housed in the Lower Nursery, where eight separate environments are maintained for growing orchids. These range from high temperature/high humidity regimes for growing tropical species, to low temperature environments for growing temperate and high altitude species, with varying degrees of temperature, light and humidity in-between. The plants are cared for by a team of four members of staff, together with a horticultural student and occasionally international trainees.

We attempt to grow as wide a range of genera and species as possible, to illustrate the extraordinary diversity of the family. Of course we can only hope to grow a small representation of the huge number that exist in the wild, so we must be careful that those we do find space for are fulfilling a useful purpose and are not simply grown 'for the sake of it'.

Each plant is labelled with an accession number (a ten figure number) which is recorded on the Kew computer together with data such as the donor or collector, its country of origin and range, flowering time, habit of growth, etc. This data is easily retrieved and is of use both to botanists' and horticulturists' research programmes. A second computer, donated to the Orchid Unit by the Sainsbury Trust, monitors environmental conditions within the eight orchid zones in the nursery; controlling variables such as temperature, ventilation, relative humidity and air movement. Finer details such as feeding, watering, potting and propagation are dealt with by an enthusiastic team of staff and students.

Source of material

Plant material is received from a variety of sources. Some is collected, under licence or by permit, from natural habitats. Such material is only encouraged or requested from knowledgeable collectors, for example from botanic gardens like Kew or residents of the country in question who know exactly what they are collecting and who respect the need for conservation. The most useful state in which material is received is as seed. This may be easily collected from the wild, together with valuable

Table 1. *A selection of orchids of particular interest in the Living Collection at Kew.*

Purpose	Plant
Display: in the Princess of Wales Conservatory	*Cattleya bowringiana* Veitch
	Disa tripetaloides (L.f.) N.E. Br. and *Disa uniflora* Berg. and their hybrid *D x kewensis*
	Hybrids of the genera *Cymbidium* Sw. and *Paphiopedilum* Pfitz.
	Dendrobium bigibbum Lindl. *Dendrobium nobile* Lindl. and its many hybrids *Pleione praecox* (Sm.) D. Don *Stanhopea wardii* Lodd ex Lindl.
Research:	Large collection of species of the genera *Paphiopedilum* Pfitz. and *Cymbidium* Sw. used in the production of recent monographs.
	Representatives of the orchid species from Borneo, East Africa and the Solomon Islands for use in forthcoming floras of those areas.
	The genus *Oncidium* Sw. – research into aspects of orchid pollination and seed germination.
Conservation: of species endangered in the wild	Very wide collection of European terrestrial orchids grown and propagated with symbiotic root fungi. Plants that will be hopefully re-introduced to the wild include: *Cypripedium* *calceolus* L., *Orchis morio* L., *Orchis laxiflora* Lam., *Ophrys* *apifera* Huds. and *Serapias lingua* L. (recently flowered from seed sown and grown symbiotically).
	Pollinations are specially made to generate seed for distribution of the rarer species of *Paphiopedilum* Pfitz. For example, *P. rothschildianum* (Reichb.f.) Pfitz, and *P.* *spicerianum* (Reichb.f.) Pfitz.
	Currently bulking-up plants and seed of the endangered *Sarcochilus fitzgeraldii* FvM. for distribution. Similarly, *Epidendrum ilense* Dodson – thought to be extinct but recently rediscovered in the wild; and *Dendrobium* *spectatissimum* Reichb.f. – endangered in the wild.

field data, without threatening wild populations, and be sent back to Kew to be germinated and weaned by the Micropropagation Unit. This obviously takes longer to produce flowering-sized plants than receiving plants directly, but often the result is greater numbers of healthier, more vigorous orchids.

We also receive plants from other botanic gardens, from nurserymen, and from interested amateur orchid growers who are keen to be involved in conservation of orchid species. Occasionally we have been able to accept large collections from growers who, for various reasons, are no longer able to keep them (for example, Lady Sainsbury, David Clulow and Barbara Everard).

Sadly we receive a large number of orchids which have been brought into the country illegally and consequently have been confiscated by H.M. Customs. In this way, we received four sizeable consignments in 1986/7. Where it is not possible to return these plants to their country of origin they are brought to Kew where they are tended until decisions can be taken on their future. Large consignments pose problems, though every effort is made to find suitable 'homes' for surplus plants in other collections.

Uses of the collection

This large and varied collection of living orchids can aid conservation in several ways, both directly and indirectly (Table 1). It is used by taxonomists, cytologists, geneticists and others to further research into the family; the revision of genera, the writing of floras and many other aspects of botanical science can all be aided by the use of living material. The more we know about these plants the easier it should be to protect them. In some cases, the Unit has carried out its own research. For example, in conjunction with Harriet Muir's work for the Sainsbury Orchid Project on symbiotic germination of European orchids, we have successfully weaned and grown on (some now into their second year) several species of *Orchis* both with and without mycorhizal fungi, experimenting with composts and with temperature/ humidity balance using a dew-point cabinet.

Each year there are many visitors to the Orchid Unit, ranging from professional scientists through Orchid Societies to parties of school children. All these people will have gained something from their visit and education of the public, especially of school children, may well be one of the keys to future awareness for the need for conservation. Still in the field of education, I have mentioned earlier that the Orchid Unit has a horticultural student working with the collection (for three to six months at a time as part of their Kew Diploma Course). Ex-Kew students are scattered throughout the globe, for example in Bermuda, Singapore, New Zealand, Chile, Canada, the Solomon Islands, Japan; many of them helping the causes of conservation in a wide variety of jobs. International trainees, often from recently developed countries, also come to work in the Orchid Unit for anything from a day to six months. They are able to take newly acquired skills back to their own countries and pass them on to other students, and so the chain of potential conservationists goes on.

By keeping a 'stock collection' of plants, the Orchid Unit could be seen in some ways as a kind of museum where orchids in cultivation at Kew are preserved as living botanical specimens for the future. However, the situation must be more dynamic if it is to actively work towards the conservation of orchid species in their natural habitat. The preservation of the actual habitat is generally an economic problem and as such is little influenced by the work of orchid growers. There will always be a market for rare and unusual orchids; this market in the past has been supplied by the collection of plants from the wild, often resulting in the near extinction of already rare orchids. The main alternative that the Orchid Unit advocates is through propagation and distribution of nursery grown plants and seedlings, to help to reduce the pressure caused by over collection in wild populations. Recently therefore we have begun to produce a modest distribution list, to try to distribute orchid species to other botanic gardens, to nurserymen and to established amateur growers, in the form of plants, seedlings, flasks of seedlings, and even pollen. It is our hope that the recipients will also propagate and distribute their plants, and in this way remove the need to collect plants from their natural habitat. It is well-known that an established nursery-grown seedling stands a much better chance of survival in cultivation than does a collected plant.

Our plans for the future? Firstly, to strengthen links with other Botanic Gardens' orchid collections – for example, in Europe, America, China and the Tropics. Secondly, to strengthen links with nurserymen and orchid growers, to encourage communication and exchange of plants between them and Kew and also between each other. Thirdly, to encourage the use of seed when collecting from the wild, and lastly to continue to try to inform the public of the diversity and beauty of the orchid family coupled with the dangers of habitat destruction and over-collection, using the facilities offered by Kew's new tropical display glasshouse, The Princess of Wales Conservatory.

Acknowledgements
Thanks to Sandra Bell and Bert Klein for providing the details for Table 1.

Import and export of orchids and the law

Introduction

The growth and development of most scientific and artistic disciplines can be traced back through history and spans centuries or even millenia. The development of conservation, a subject largely based on scientific as well as ethical principles is still in its infancy, having only commanded serious academic attention in recent decades. There are now several international wildlife treaties and conventions and fortunately for both professionals and non-professionals interested in the subject the current status and legal background have been brought together by Lyster (1985), to whom much of the foregoing information is attributable.

Control on international trade in wildlife, its products and derivatives is not a recent concept. Initial public demand for such controls was made as early as 1911 by the Swiss conservationist, Paul Sarasen, who claimed that the vogue for plumed hats was having a serious effect on populations of wild birds. Sarasen was one of the most influential figures behind the establishment of the Consultative Commission for the International Protection of Nature at Berne in 1913, which had delegates from seventeen European countries. The progress of the Commission was halted by the outbreak of War and it was not until the late 1940s when the foundation of a similar international body was under discussion, that the Commission had any legal existence. Furthermore it set a precedent, being the first intergovernmental agency concerned with nature protection (Boardman 1981).

The International Union for the Protection of Nature, later becoming the International Union for Conservation of Nature and Natural Resources (IUCN), addressed the problems of international trade in wildlife. At the eighth meeting of the General Assembly of IUCN held in Nairobi in 1963, there was a call for 'an international convention on regulations of export, transit and import of rare or threatened wildlife species or their skins and trophies'.

International conventions

The IUCN initiative was taken in a climate of increased public awareness for the need to protect species occurring alongside Man on this planet, and ten years later in 1973, the Convention on International Trade in Endangered Species of Wild Fauna and Flora (CITES) was concluded in Washington, D.C. and signed by twenty one States. The Convention did not enter into force until July 1975, this being ninety days after the tenth signatory had deposited an instrument of ratification. Over the last decade the Convention has greatly expanded and at the time of writing ninety two States are party to the Convention.

Although several international conventions concerned with the protection of wildlife are now in force, CITES has proved to be by far the most successful. This success is attributable to a number of factors. The basic principles of CITES – no international commercial trade, with a few exceptions, in endangered species and a carefully regulated trade in species which are not yet endangered but may become so – have proved widely acceptable. Most States recognise that the regulation of international trade requires the cooperation of both producer and consumer countries. Furthermore, the administrative structure set up by CITES to oversee its permit system ensures that the Convention is better enforced than many other wildlife treaties.

CITES appendices

Species controlled by the Convention are listed on three Appendices. Species not yet threatened with extinction but which may become so unless trade is regulated are listed Appendix II. These species may be traded under licence as long as such trade is not considered by a Scientific Authority of the exporting country to be detrimental to the survival of the species. Those species of plants and animals considered to be threatened with extinction are listed on Appendix I. International trade in these species is generally prohibited, but in exceptional circumstances may be allowed if the Parties concerned consider such trade is not detrimental to the survival of the species in question and is not for primarily commercial purposes. Specimens of Appendix I species which are captive bred or artificially propagated are treated as if they are in Appendix II. Appendix III provides a mechanism whereby a party having domestic legislation controlling export of species not already listed in Appendix I or Appendix II can call upon the help of other Parties in the enforcement of this legislation.

Operation of the Convention

The administrative structure established by CITES can be grouped into three recognisable units. These are the Secretariat, the Conference of the Parties and the Management and Scientific Authorities. All parties are required to designate Management and Scientific Authorities to implement the Convention which is operated by a permit system.

In addition a Secretariat operates from Switzerland and ensures that the system is being correctly implemented internationally. Furthermore, it is a requirement that the Parties meet every two years to review the implementation of the Convention and discuss recommendations for amendments to the Appendices.

Table 1. *Addresses of institutions from whom application forms and further information can be obtained.*

1. Phytosanitary Certificates:
   Plant Health Division
   Ministry of Agriculture, Fisheries and Food
   Great Westminster House
   Horseferry Road
   London SW1P 2AE
   Tel. no. (01) 216 6174

   Plant Health Division
   Department of Agriculture and Fisheries for Scotland
   Chesser House
   500 Gorgie Road
   Edinburgh EH11 3AW
   Tel. no. (031 443 4020)

2. CITES Certificates and Permits:
   Endangered Species Branch
   Department of the Environment
   Tollgate House
   Houlton Street
   Bristol BS2 9DJ
   Tel. no. (0272) 218202

   Wildlife Licensing Section
   Department of Agriculture for Northern Ireland
   Dundonald House
   Upper Newtonards Road
   Belfast BT4 3SB
   Tel. no. (0232) 650111 ext. 642

In the U.K. the designated Management Authority is the Department of the Environment (see Table 1) and in 1976 the Secretary of State for the Environment designated the Royal Botanic Gardens, Kew as the U.K. Scientific Authority for Plants and this represents the only statutory role of the Royal Botanic Gardens. In 1981 the Nature Conservancy Council was designated as the Scientific Authority for Animals. Both Scientific Authorities advise the Management Authortiy on all applications to import and export species covered by the Convention.

CITES permits and certificates

The Management Authority of each Party controls the issuing of permits to import and export plants, and the Convention requires that each consignment of specimens is accompanied by a seperate permit or certificate.

Table 2. *Appendix I listing of* Orchidaceae.

Species	Distribution
Cattleya skinneri Batem.	Belize through Central America to Venezuela[a]
Cattleya trianaei Linden & Reichb.f.	Colombia
Didiciea cunninghamii King & Prain	Sikkim and Uttar Pradesh
Laelia jongheana Reichb.f.	Brazil
Laelia lobata Veitch	Brazil
Lycaste virginalis Hook. var. *alba* Cockerell	Mexico and Honduras[a]
Paphiopedilum druryi (Beddome) Stein	India
Peristeria elata Hook.	Costa Rica to Venezuela
Renanthera imschootiana Rolfe	Burma, Manipur and Nagaland
Vanda coerulea Griff. ex Lindl.	Burma, Thailand, Manipur, Meghalaya, Nagaland

[a]National flower

Plants which have been artificially propagated in the country of origin should be accompanied by a CITES certificate. The subsequent movement of these plants across any border should be accompanied by this certificate and this negates the need to re-apply for permits from each Party State through which the consignment passes. The advantages to hybrid orchid traders is quite clear. The need for a lot of unnecessary paper work is removed. Controls are obviously stricter for plants collected in the wild and in these instances Management Authorities may request the importer to obtain an export permit from the country of origin, prior to allowing the issue of an import permit.

Trade in all species of the Orchidaceae is controlled by CITES. Ten species are listed on Appendix I and these are given with their distributions in Table 2. At a recent meeting of the CITES Plant Working Group, general agreement was expressed regarding the proposal suggested by the Guatemalans to downlist *Cattleya kinneri* from Appendix I to Appendix II, although this was not passed as a formal amendment at Ottawa in 1987.

The remaining species are all listed on Appendix II and therefore all international trade in them is monitored. Each Party is required to produce an annual report on their trade in CITES species and should a situation arise where the level of trade is considered to be detrimental to either a population of the species or the existence of a species in the wild, then the Management Authority is empowered to refuse the issue of permits for such species. Ironically, several species listed on Appendix II are in much greater danger than their counterparts on Appendix I. A classic example is given by the rare *Paphiopedilum sanderianum* thought to the extinct until its recent rediscovery in Sarawak (Alexander 1984; Argent 1984). Sadly much of the population is thought to have been exploited by a commercial collector and it is an unhappy reflection of present times that accurate scientific data on rare plants can no longer be published without fear for the future of the population of the plants in question.

A standard model for CITES permits and certificates was approved at the New Delhi Conference of the Parties in 1981 in an attempt to foil forgeries. It was recommended that Parties should adapt their permits as closely as possible to the standard model.

Phytosanitary certificates

In addition to CITES certificates and permits any plant entering the U.K. must be accompanied by a Phytosanitary Certificate. The orders imposing these restrictions are the Import and Export (Plant Health) (Great Britian) Order 1980 and the Import and Export of Trees, Wood and Bark (Health) (Great Britain) Order 1980, as amended in 1983 (Anon 1983). Both orders cover imports into England, Scotland and Wales and separate legislation covers the import of plants and their products into Northern Ireland, Channel Islands and the Isle of Man. Phytosanitary regulations are enforced by the Ministry of Agriculture, Fisheries and Food (see Table 1). Phytosanitary Certificates must be issued by the plant protection service of the country of export. Certificates issued by local government departments are not acceptable. As with CITES permits and certificates, imports from other European Community (EC) member states are less restricted than those coming from third countries.

Passenger baggage concessions

The Import and Export Order 1980 allows for the landing of certain plant groups in Great Britain without the need for a Phytosanitary document and these concessions are listed in Table 3.

The importation of cut flowers of orchid species and hybrids is permissible under CITES and Plant Health Regulations, provided that the plants were artificially propagated. For example, a bunch of *Vanda* hybrids purchased in Malaysia could be imported into Britain without either Phytosanitary certificates or CITES Permits. Another exception to the requirement for certificates and permits is the importation of fruits of artificially propagated species of *Vanilla*. Similarly, tissue cultures, flasked seedling cultures, seeds and pollen are also exempted.

In both instances it would be preferable if a certificate declaring that the plants were artificially raised could be obtained at the time of purchase. This requires no more effort on the part of the nurseryman or shop assistant and the purchaser than asking for a receipt.

Table 3. *Passenger baggage concessions.*

The following goods can be imported into the U.K. as part of passenger baggage concessions from the following countries:

Countries within the Euro–Mediterranean area
> up to 2 kg of tubers bulbs and corms;
> up to 5 plants or parts of plants;
> a small bunch of cut flowers; and
> up to 2 kg of fruit and vegetables (but not potatoes)

There are no restrictions on imports of flower seeds from any country.

N.B. This concession does not apply to the following:

Potatoes, forest trees, fruit tree material, chrysanthemums, vine plants, plants or seeds of *Beta* species and cut gladioli from Malta.

Countries outside the Euro–Mediterranean area

> There are no concessions for any plants, their materials or products which all require a Phytosanitary Certificate. However, unrestricted seed, raw fruit and vegetables, and cut flowers may be imported in the usual way.

U.K. enforcement of CITES

In the U.K. enforcement of the Convention is undertaken by the Department of the Environment in conjunction with H.M. Customs and Excise. Conservation controls on the import and export of CITES species are applied in the U.K. under the provision of the European Community CITES Regulations. Any Party may take stricter controls than those laid down in the Regulations and the U.K. has elected to do this. One example is the U.K. requirement for import permits to be supplied for Appendix II species even though the Convention only requires the production of export permits.

Furthermore the U.K. controls trade in non-regulation species of wildlife under the Endangered Species (Import and Export) Act 1976. This includes those species covered by the Convention on the Conservation of European Wildlife and Natural Habitats (commonly known as the 'Bern Convention'), and the controls apply equally to trade with EC and non-EC States.

Illegal importers may be charged under several sections of the law and one case was brought to court in 1986 by H.M. Customs and Excise. A nurseryman in south west England was prosectued under the Customs and Excise Management Act and pleaded guilty to six specimen charges relating to the import of orchids and *Cyclamen* species without the appropriate CITES documentation. The total fine was £1,800 and the orchids involved included *Orchis militaris, Orchis simia* and *Himantoglossum hircinum*.

Conclusion

During the twelve years that CITES has been in force, much progress has been made. Most of the nations who trade in wildlife have ratified the Convention with two important exceptions as far as plant trade is concerned. These are Mexico and Turkey. Trade in Appendix I species has been described as sporadic and international trade in Appendix II species is now carefully regulated. By having a permanent Secretariat and many obligations that must be filled, CITES remains an active treaty and hopefully will continue to be an example to other treaties concerned with the international conservation of wildlife.

References

Alexander, C. (1984). *Paphiopedilum sanderianum. Kew Mag.* 1, 3–6.

Anon. (1983). *Plant Health Import Legislation Guide for Importers.* London: Ministry of Agriculture, Fisheries and Food.

Argent, G. (1984). *Paphiopedilum sanderianum* (Rchb. f.) Stein flowering in cultivation. *Orchid Rev.*, **92**, 208.

Boardman, R. (1981). *International Organization and the Conservation of Nature.* London: Macmillan Press.

Lyster, S. (1985). *International Wildlife Law.* Cambridge: Grotius Press.

INDEX

anamorphic forms, of Basidomycete endophytes, 62–3
asymbiotic, seed germination, 31–8
atmosphere composition, and seedling/protocorm growth, 73–84
auxins, in tissue culture media, 33, 92–3

Basidomycete fungi, and specificity for germination, 61–3
Brazilian rainforest, conservation project, 153–8
bud development, from apical meristem, 95

C3 carbon fixation, in protocorms/seedlings, 83–4
callus development, from explants for tissue culture, 90, 95
carbon dioxide, concentration and protocorm/seedling growth, 75–84
cell number, in *Cattleya aurantiaca* seed, 21–5
cell wall, degradation during fungal infection, 61
CITES, Convention on International Trade in Endangered Species of wild fauna and flora, 163–9
species listed under Appendix I, 166
classification of Basidomycete fungi, 61–3
CO2 *see* carbon dioxide
compatibility, of fungus for seed germination, 39–56
recognition by host, 66–7
see also specificity
Crassulacean acid metabilism (CAM), in advanced epiphytic genera, 83
cryopreservation
of pollen, 11–14
of seeds, 17
cytokinins, in tissue culture media, 92–3

decontamination, of explants for tissue culture, 90–1

see also surface-sterilization
distribution, and species rarity, 141–5
dormant states
in plant life-history, 106, 108, 118–19
in plant recruitment, 122–6
dry weight
of growing protocorms/seedlings, 68–9, 82
of plants, 131, 134–5

equilibrium moisture content
of pollen 6, 8
of seeds, 18
ethylene, concentration and protocorm/ seedling growth, 76–8

flower, frequency per inflorescence, 135, 137–8
flowering
prediction of in *Ophrys apifera*, 127–39
time from emergence to, 104–8
fungal
(in)compatibility in seed germination, 39–56, 57–71
infection patterns in protocorms and roots, 58–60
isolates for seed germination, 39–56, 61–3

germination
asymbiotic method for seeds, 31–8
in pollen of European terrestrial species, 4–8
in seeds of European terrestrial species, 32–7, 39–56, 59–69
symbiotic method for seeds, 39–56, 57–69
see also viability

half-life, plant life-span, 104–5, 108
host-fungus relationship
infection pattern, 58–9, 66
in mycorrhizal systems, 57–71
seedling nutrition and, 66